The Real Starting-Point for Our World

The Metaphysical Implications of Quantum Discontinuity

Garry Seabrook

Copyright © 2024 by Garry Seabrook

www.garryseabrook.com

The author asserts the moral right to be identified as the author of this work

All rights reserved. No part of this book may be reproduced in any form or by any electronic or mechanical means, including information storage and retrieval systems, without permission in writing from the publisher, except by a reviewer who may quote brief passages in a review.

Paperback ISBN: 978-0-6453525-3-5

Ebook ISBN: 978-0-6453525-4-2

Typesetting, cover design by Garry Seabrook

Cover photograph: faraz ahanin: https://www.pexels.com/photo/pattern-on-desert-sand-18325576/

Printed by Ingram Spark, Ingram Content Group, 1 Ingram Blvd., La Vergne, TN 37086 www.ingramspark.com

Contents

Dedication	IV
Introduction	V
1. The Problem Begins in Western Metaphysics	1
2. The Quantum Mystery and Particle-Wave Duality	15
3. The Application of Noncontradiction is No Longer an a Priori Certainty	25
4. Quantum Discontinuity is an Ontological Problem	38
5. The Critical Ontology of Nicolai Hartmann	49
6. Kant's Unwitting Error	61
7. Hegel Made the Same Unwitting Error	78
8. Again, the Problem Starts in Western Metaphysics	94
Endnotes	106
Bibliography	110

for Willy

INTRODUCTION

This book, as the name suggests, deals with the starting-point for our world. It doesn't talk about the Big Bang, it doesn't mention it beyond this introduction, and it actually makes no judgment at all regarding the accuracy of this theory in being able to describe the origin of our universe.

The book focuses instead on our understanding of contradiction, specifically the way we apply the principle of noncontradiction as a fundamental law in our world, as the first law of logic, and as the effective starting-point for all knowledge in and about our world. It argues that we've always mixed-up the idea of non-contradiction as a simple truism and the application of this truism as a fundamental law in our world. The discovery of quantum interaction has brought into question the straightforward way in which we've always applied this law.

The simplest explanation for the existence of quantum interaction is the emergence of causality from no-causality (i.e., randomness). This is recognised by leading Commentators on quantum physics.[1]

The problem is, nobody has yet been able to work-out how to reconcile this emergence with our metaphysical beliefs about the world. The difficulty lies ultimately with our metaphysical starting-point, that is to say, our application of the truism of noncontradiction as a fundamental ontological law.

The key to solving the quantum mystery is to understand the implication of this discovery with regard to our application of the law of noncontradiction and particularly its role as the starting-point for all knowledge. Mathematical descriptions, such as the Big Bang, are necessarily dependent on the application of this law for their ultimate connection to our world. Put simply, it would be impossible to know anything about our world without first having a clear understanding of how this law initially applies.[2] Because of its obvious certainty as a logical truism, the straightforward application of this law has always been taken for granted.

Philosophers have questioned the role of this truism as the first law of logic, but they've never seriously doubted its straightforward application as an unavoidable truism in our world. To do so is considered logical heresy! Again, this is because we conflate the application of noncontradiction with the idea of it as a simple truism. Efforts to develop a quantum logic, for example, have attempted to reassess the way we apply noncontradiction based on the behaviour of quantum objects, but they've never really questioned the self-evident certainty of this truism itself and the validity of its initial application to these objects.[3] Even the supposed existence of such objects ultimately owes its origin to this application.

The very fact we presuppose that the world can be described mathematically and logically is first due to our initial application of the truism of noncontradiction.

The approach taken here is metaphysical, although it is also critical of Western metaphysics and contemporary mainstream philosophy in general. This criticism is again aimed at the way mainstream philosophy commonly takes the straightforward application of the truism of noncontradiction for granted. The traditional approach of Western metaphysics is a priori, that is, it strives for certain knowledge about the real world starting from first principles, particularly the law of noncontradiction. Such first principles are naturally presupposed not to require reference back to our common experience of the world. That's what defines them as a priori principles.

Most contemporary philosophers set aside the need for an ultimate starting-point for everything, satisfied in the validity of mathematics and also logic to be able to access the metaphysical foundations of our world. Buoyed, perhaps understandably, by the success of modern physics to be able to describe mathematically the physical world, the metaphysical structure of our world is also almost universally taken to be entirely describable mathematically. The priority of contemporary philosophy is to discern the logical structure of our world, in the hope, essentially, of emulating the success of modern physics. Because its metaphysical structure is taken to be entirely logical, the need to clearly define an ultimate starting-point is assumed to be unnecessary. This assumption, again, takes for granted the self-evident certainty of the truism of noncontradiction and its necessary application in our world.

As a consequence of presupposing the metaphysical structure of our world to be entirely describable via mathematics and logic, contemporary thinking has effectively extricated humankind from its deductive calculations about that structure. In other words, we presuppose ourselves simply to be products of that metaphysical structure and, more significantly, as passive observers when it comes to deductively describing that structure. Even though this way of understanding our position in the world may be essentially correct, it does create a problem when it comes to discerning the ultimate starting-point for everything, not because of some quaint desire to place human cognition at the centre of the universe, but because of the simple fact that such a starting-point would also have to serve as the initial starting-point for all knowledge in and about our world.

Again, this is not really a problem whilst ever we take the law of noncontradiction merely as a logical truism because this truism is then able to serve as the a priori starting-point for all our knowledge. This is how philosophers have traditionally approached the metaphysical structure of our world, Kant and Hegel, for example, and David Hume, by taking the law of noncontradiction as something we could not possibly be deceived about. Because of its obvious certainty, we've always presupposed it unnecessary to refer the application of noncontradiction back to our experience of the world. This is the way we've always commonly understood the application of this law in our world.

While it may be convenient to understand the world in this way, it distorts our judgment when it comes to understanding the meaning of quantum interaction, and particularly the significance of its randomness, its discontinuity in space and time. This is because, what we may in fact be looking at with the relationship between this spatiotemporal discontinuity and the continuous causal structure of our world, is the very starting-point itself for our world, at least, that is, how that starting-point appears to us from within and as part of the same spatiotemporal world.

By approaching our analyses of quantum interaction simply as passive observers, we effectively overlook the potential significance of this discovery to knowledge itself and specifically the implications of this discovery on the way knowledge initially connects to our world. We're unable to realise that the discovery of a real contrary relationship at the very spatiotemporal limit of our observable world must bring into question the a priori status of the law of noncontradiction, and specifically how we've always applied this law simply as a logical truism in our world. The mere possibility of a real contrary relationship physically existing prior to the application of the truism of noncontradiction, potentially even defining the starting-point itself for our world, must render moot the a priori status of this law. It's the discovery at the very limit of our observable world of a contrary relationship between spatiotemporal discontinuity-continuity that makes such a possibility plausible.

Again, the problem is, we've always mixed-up the idea of noncontradiction as a self-evident truism with its real application in our world. It's this application that codifies noncontradiction as a fundamental ontological law, not merely its truism, and it's this application that can no longer be understood simply as an a priori certainty following the discovery of quantum discontinuity and particularly its contrary relationship in space and time with the continuous causal structure of our physical world.

The simplest explanation for this relationship is the emergence of causality from randomness, in other words, the emergence of causality from no-causality. The reason this is not commonly accepted as the explanation for the mystery surrounding quantum interaction is ultimately because of our inability to reconcile it with our application of the truism of noncontradiction. We continue instead to try and formulate a mathematical solution to this mystery that can account for our observations of quantum objects, and that continues to presuppose the application of the truism of noncontradiction to such objects.

The crucial point to understand here is that, if the emergence of causality from no-causality represents the real physical starting-point for our world, such a starting-point would have to precede our initial application of the truism of noncontradiction. It's the application of noncontradiction, not its status as a logical truism, that actually defines it as a fundamental principle and first law of logic in our world. This physical starting-point would literally have to come before everything in our world, including all possible knowledge and the application of any laws, even the first law of logic.

We've always taken it for granted that, because nothing can possibly contradict itself, the truism of noncontradiction can provide an a priori starting-point for logical thinking, and thus also, for all knowledge in our world. With the emergence of causality from no-causality, clearly, we're not talking about a contradiction, but a real contrary relationship physically existing prior to the initial application of the law of noncontradiction. A contradiction will always be a contradiction, but how this truism initially applies as a fundamental law in our world must be determined by such a real starting-point and not simply its status as a logical truism.

By presupposing the application of noncontradiction as an a priori truism, we effectively pre-define the starting-point for our world. It doesn't matter even if we conclude that such a starting-point is ultimately unknowable, we've already presupposed it as necessarily being dictated by the truism of noncontradiction. That is to say, we've logically determined in advance that any such starting-point must, at the very least, be pre-defined by the mutual exclusion of contrary relationships. In the case of the emergence of causality from no-causality, to be more accurate, the relationship between spatiotemporal continuity-discontinuity that defines the limit of the observable world from our location within and as part of that world, the necessity for such mutual exclusion only really captures half the story.

Because this contrary relationship effectively defines the absolute starting-point for everything in our world, even preceding any possible application of the law of noncontradiction, it can also be understood to define the way noncontradiction initially applies as a fundamental law in our world. In other words, our initial application of the law of noncontradiction is determined not merely by the necessary mutual exclusion of contrary relationships, as has been always presupposed by the idea of this law as a logical truism, but specifically by the simultaneous mutual exclusion and joint completion of the relationship between quantum discontinuity and the continuous causal structure of the physical world that has been discovered to define the measurable limit of our world—that is to say, the inherent complementarity in space and time of this relationship.

It's the realisation of this fact that represents the true significance of the discovery of quantum interaction, and it's this realisation that makes all the difference, not just for solving the mystery surrounding this discovery, but ultimately for our metaphysical understanding of the world.

One

The Problem Begins in Western Metaphysics

The main reason we struggle to understand the meaning of quantum interaction is because we naturally assume it to be a problem for science, physics, to be exact, when really, it's a philosophical question and specifically metaphysics. The simple reason it's not a physics problem is because it concerns the starting-point for our world.

> But, surely, that's still a physics problem. We're still talking about the physical world.

Well, that's the accepted way of thinking, and a mistake, I believe, especially when it comes to trying to figure out the ultimate starting-point for everything. As I'm going to explain, this is most likely what we're looking at with quantum interaction and such things as particle-wave duality. I'm going to talk more about this duality later; suffice for now to point out that this duality marks the limit of what we can observe and physically measure in our world.

Actually, let's be more precise from the outset. The limit of what we can observe and measure in the physical world is marked by the spatiotemporal discontinuity of quantum interaction. What I mean by this is simply that quantum interaction always occurs randomly (i.e., discontinuously) in space and time. This discontinuity contrasts with what we otherwise understand to be the spatiotemporally continuous causal structure of our world.

Quantum interaction just refers to the way the quantum world interacts with the physical world, and as I just said, this always happens discontinuously in space and time. Sometimes, I'll refer to this interaction in terms of quantum events. This is because we literally always observe this discontinuous interaction as tiny little random flashes of light.

How we understand quantum interaction really all hinges on our interpretation of the relationship between spatiotemporal discontinuity-continuity that ultimately defines the limit of what we can measure in the physical world.

By quantum objects, I just mean the quantum things, the particles and waves, that we assume to exist based on our quantum experiments and our observations of quantum interaction. Again, I'll talk more about quantum objects later. Suffice for now to point out that our entire understanding of these objects and the way they seem to behave, particle-wave duality, for example, can be traced back to this relationship between spatiotemporal randomness (i.e., discontinuity) and the continuous causal structure of our world. The key, really, is to understand the reason for this randomness and particularly the significance of its discontinuity in space and time.

For reasons which should become clear, there is a distinction that needs to be drawn between what we can measure and what might lie beyond this observable discontinuity. Quantum mechanics does a thorough job of describing and predicting the effects of quantum interaction in our world, but I argue that it's not suited to working out the metaphysical reason behind this interaction.

> So, you're trying to say that quantum mechanics is wrong!

Not at all! Let's be clear from the beginning. What I'm talking about really has very little to do with quantum mechanics. By quantum mechanics I just mean the mathematics used to describe quantum interaction. The physical effect on our world of this interaction is certainly a problem for physics, and this can be understood mathematically. The mystery behind it, on the other hand, is a metaphysical problem, and it needs to be addressed metaphysically. Can you remember what I mean by quantum discontinuity?

> Isn't it the randomness in space and time of quantum interaction? We always observe quantum interaction just as tiny little random flashes of light.

Even within metaphysics there is a distinction to be recognised between the study of how we know the metaphysical world, referred to as epistemology, and the study of the nature of the metaphysical world itself, or ontology. We talk about this ontic world in terms of Being and existence. Ontology is the study of the ontic structure of existence. We also refer to this ontic structure as the 'real' structure of the world, in contrast to the world we commonly experience every day, and what philosophers take as representing only an 'appearance' of this real world. It's this idea of an ontic structure that we're really talking about with the quantum mystery, and, as I hope will soon become clear, the solution to it needs to be understood ontologically.

For several hundred years, Western metaphysics has preoccupied itself mainly with epistemology, or the question of how we can know the real world. This preoccupation with the question of knowledge is also part of the problem when it comes to understanding what's behind quantum mysteries such as particle-wave duality. We presume, mistakenly, I believe, that we need to work out *how* we can know the real world before we can figure out *why* our world actually appears the way that it does.

> But surely, we need to figure out how we can know the structure of the real world before we can understand why it's the way that it is.

That's a very common assumption. The point is that we have to be clear first about the starting-point itself before we can work out anything else, even and including the epistemological question of how we know the real world. We're talking about the starting-point for literally everything in our world, including all possible knowledge. It's something that almost certainly impacts upon our most basic assumptions about the world, and the way we think, not to mention our very conception of knowledge itself. It's not even a chicken and egg scenario between epistemology and ontology, once the age-old starting-point for everything has been brought into question, and that's what the discovery of quantum discontinuity has effectively done.

We've always taken our logical starting-point for granted because we've always assumed it to be guaranteed by the law of non-contradiction. We've grown complacent about this starting-point because of the seemingly self-evident nature of the application of this truism in our world. Both our scientific and philosophical approaches to the problem of quantum discontinuity naturally presuppose the validity of the application of this truism to the quantum realm. Instead of questioning this application, we doggedly cling to the supposed certainty of mathematics and logic in trying to solve the mystery of quantum discontinuity. This way of thinking is a mistake, I believe, and it is the main reason why we fail to grasp the real meaning of this discovery.

I don't understand.

Don't worry, we're getting a little ahead of ourselves. Suffice for the moment to assert that the ontic meaning of quantum discontinuity is something that needs to be addressed first ontologically, not epistemologically, logically or mathematically.

The famous debate between Einstein and Neils Bohr, for example, boiled down to an epistemological argument, and the question of whether we could know the real cause of quantum interaction, specifically, whether we could rely on mathematics to describe the quantum world beyond what could be measured. Einstein believed we could, while Bohr thought, maybe we couldn't.

Einstein and Bohr were talking as much about metaphysics as they were about physics. Their respective arguments can be understood to reflect the epistemological positions of the philosophers Hegel and Kant.[1] I'll talk more about these guys later, but basically Hegel believed we could access the real structure of the world using logic and mathematics while Kant thought this might not be possible, at least, not with any certainty. There's a simple reason why both these arguments actually miss the point when it comes to the metaphysical significance of quantum discontinuity. Einstein and Bohr, like Hegel and Kant before them, simply took as self-evident the initial application of the truism of noncontradiction to our world.

Of course, today the overwhelming majority of scientists and philosophers go along with Einstein and Hegel's way of thinking, that we can describe the real structure of the world through mathematics and logic, even beyond what can be experienced and measured. Contemporary philosophers might not all agree with Hegel's ideas, but they do mostly accept this point about mathematics and logic.

Indeed, the vast majority of contemporary philosophers have come to accept mathematics and logic as basically the same in this regard, while physicists assume, almost universally, that the real structure of our world is completely describable through mathematics.

The success of modern science has resulted in a fundamental change in our metaphysical assumptions about the world, even to the extent that the traditional study of metaphysics itself has come to be considered by many a waste of time. This is understandable, I suppose, especially given the great success of mathematics in being able to describe the physical world. Nowadays, we naturally take it for granted that everything is contained physically within our spatiotemporal world of moving parts, and that it's all entirely governed by the law of efficient causality. Any randomness in our world must have its origin somewhere within this causally governed structure.

Philosophers and scientists don't even trouble themselves too much anymore with the problem of an ultimate starting-point for everything, providing the physical world continues to adhere to the law of efficient causality and, of course, the self-evident certainty of the truism of noncontradiction. Regardless of the starting-point, as long as our world is entirely logical, it should all be describable using mathematics and logic.

It's not hard to understand how the boundary between the physical and the metaphysical has become blurred. It's also not hard to see how some of the greatest minds of the last hundred years could be seduced by the idea of trying to solve what is presumed to be a profoundly complex mathematical problem behind quantum discontinuity. Einstein, for most of his life, applied his renowned grey matter to trying to figure this problem out, unsuccessfully, the so-called mathematical theory of everything, as did Stephen Hawking.

Not surprisingly, the main sticking point is the incompatibility between the causally governed and spatiotemporally continuous world that physicists have mathematically pieced together over the last several hundred years and the randomness in space and time of quantum discontinuity. It all pretty much boils down to a contradiction between the mathematics of Einstein's space-time continuum of general relativity and the spatiotemporal discontinuity at the heart of quantum mechanics.

The point I want to make is that the solution to this problem could actually be a lot simpler than we think, and it may, in the end, just boil down to a re-think of the way we apply the law of noncontradiction to this relationship.

Really! How so?

You see, the solution to the quantum mystery, the real starting-point for our world, and how we apply the idea of noncontradiction as the first law of logic could all amount to the same problem. We've always overlooked this possibility because, basically, we mix-up the application of the law of noncontradiction with the idea of it as a simple truism.

Physics, epistemology and logic all naturally take it for granted that it must be the truism itself that applies in their calculations about the real structure of our world. Again, the problem is, they mix-up this truism with its application. This may seem a subtle point, but it makes all the difference, especially when it comes to understanding the ontic structure of our world.

First, we need to appreciate that the law of noncontradiction applies, before all else, as a fundamental ontological law. In other words, it begins as a basic law of existence—I should say, it *applies* as a basic law of existence—and after this, as the first law of logic.

The truism of noncontradiction was originally identified by Aristotle as the starting-point for all knowledge.[2] The key importance of this role has been lost in modern times, partly, again, because it's taken simply to apply as a truism and because of the presumption that the ontic structure of our world can be described entirely through mathematics.

Because everything literally hinges on this initial starting-point, potentially even the accuracy of our mathematical descriptions of the physical world, we do need to be clear about its ultimate nature. It's this starting-point that has been brought into question by the discovery of quantum discontinuity. Specifically, we need to reassess the application of the truism of noncontradiction as a fundamental ontological law.

It's because of this mix-up with the law of noncontradiction that Einstein and Bohr ultimately missed the point in their epistemological debate over quantum discontinuity, and it's why we universally fail to understand what this interaction is actually telling us about the real structure of our world.

Once we understand this metaphysical error and set aside the application of this truism to the ontic structure of our world, the best explanation for the mystery surrounding quantum discontinuity becomes the emergence of causality from randomness.

The scientific definition of randomness is simply the absence of any causality. For this reason, it is perhaps more appropriate to refer to this ontic randomness merely as no-causality.[3] First there is no-causality (i.e., randomness), and then causality emerges from it. It doesn't even matter how this emergence happens. As a scientific concept, 'emergence' is defined by the fact that, ultimately, it's something that can't be explained, either mathematically, or otherwise. It just happens.[4]

When I say that the emergence of causality from no-causality represents the best explanation for the origin of quantum discontinuity, I mean that it represents the simplest plausible explanation for the existence of this discontinuity. This assertion is based on the philosophical principle known as 'Occam's razor', which states to the effect that, given a choice, the simplest explanation is normally the best.

In the case of quantum discontinuity, the simplest and most plausible explanation is the emergence of causality from no-causality.

> But, surely Einstein, Bohr, Hawking, or any of the other smart physicists or philosophers would have realised that?

Maybe not, and part of the reason is because the implications of this discontinuity run in the face of not just physics, but also, and perhaps even more importantly, our traditional understanding of metaphysics, starting with our application of the law of noncontradiction.

This really is the significance of this discovery at the very limit of the physical world. We continue to try and solve this mystery by following our age-old metaphysical assumptions, particularly about the law of noncontradiction, and also the theories of Hegel and Kant, for example, without realising that it's actually these basic metaphysical assumptions that have been brought into question because of this discovery.

If quantum discontinuity does all hinge on the emergence of causality from no-causality, then such a dynamic would have to come before literally everything in our world, including all knowledge in and about our world.

Understand, in our all-encompassing world, it's impossible by definition for there to be any knowledge other than knowledge about our world. I prefer not to use the expression 'reality' here, but rather the concept of 'our world' as referring to this all-encompassing world that we live in. As the very starting-point for literally everything, the emergence of causality from no-causality would have to precede absolutely everything in our world, including all possible knowledge.

Look at it this way, by taking the application of noncontradiction for granted, we are effectively defining already the starting-point for our world, and thus also by implication the ontic structure of our world. As what we are really talking about is the ontic structure of existence itself, it's fair to say that this structure must also double as the effective starting-point for literally everything. We presuppose that this ontic structure can't possibly be random and causally governed at the same time because such a relationship would supposedly contravene the truism of noncontradiction. In doing this, however, we fail to grasp that, if this initial starting-point for everything actually derives from the physical emergence of causality from no-causality, then this real relationship would have to exist already, before this truism ever applied.

By taking for granted the application of this truism, we naturally presuppose the need for a choice between causality and no-causality even at the ontic level. We refer to this causal structure of existence in terms of a causal ontology. The idea of a causal ontology began to be taken for granted following Galileo, and certainly after Newton. Kant, Hegel, and even David Hume, assumed the existence of such an ontology. What these philosophers questioned and tried to work out was the epistemological problem of how we could know this causally governed ontic structure.

In doing so, they naturally presupposed the logical starting-point for their analyses to be the unavoidable truism of noncontradiction, essentially taking for granted its initial application to our world. It's not its truism that defines the law of noncontradiction in our world, but its necessary application as a fundamental law. This truism can't function as a law in our world until it has been applied as such. Until now, this point doesn't seem to have been sufficiently appreciated.

When it comes to the law of noncontradiction, the truism itself is, of course, always a truism: a contradiction is a contradiction. However, the application of this truism to our world, that is to say, even as a fundamental ontological law of existence, must come after the emergence of causality from no-causality as the ultimate starting-point for our world. Even Einstein, Bohr or Hawking don't seem to have picked up on this point. We always assume the truism of noncontradiction is an absolute certainty that must ultimately apply even with regard to quantum interaction. Some philosophers may question the role of this truism as the first law of logic, but they never seriously doubt the ultimate validity of its application in our world.

No-causality-causality does not simply contradict in the quantum mystery because this relationship literally comes before the application of the law of noncontradiction. This is why, as I said, the quantum mystery, the ontic structure of our world, and the application of the law of noncontradiction all amount to the same problem. As the starting-point for absolutely everything in our world, no-causality-causality is not merely contradictory—or mutually exclusive in the sense that these two properties can't both co-exist in the same object simultaneously—but they must also combine together to jointly complete the starting-point of our world. In other words, the relationship between no-causality-causality can be taken to define the ontic starting-point of existence itself.

It's significant that, when Aristotle originally introduced the principle of noncontradiction as the first law of logic, more than two and a half thousand years ago, he pointed out that it was commonly understood by philosophers at the time that the starting-point for our world was the contrary relationship between continuity and discontinuity.[5] In other words, this was how philosophers in Aristotle's day interpreted their experience of our world. He wrote this in a text he originally referred to as *First Philosophy*, but which has come to be known today as the *Metaphysics*. Aristotle then preceded, basically, to argue that this contrary starting-point had to be governed ultimately by the truism of noncontradiction. This in a nutshell led to the ontological mix-up that philosophers and now physicists have been struggling with ever since. Of course, Aristotle had no way of knowing then about the existence of quantum discontinuity or Einstein's space-time continuum.

Again, applying Occam's razor, the simplest and most plausible explanation for the discovery of quantum discontinuity at the very limit of the physical world is the emergence of causality from no-causality. Remember how I defined quantum discontinuity?

> Doesn't it refer to the randomness in space and time of quantum interaction?

Think about it, if such randomness (i.e., no-causality) does exist as a jointly completing part of the ontic starting-point for our world, then it would have to appear to us as spatiotemporally discontinuous, that is to say, from our location within and as part of the same world, and as part of the same otherwise spatiotemporally continuous causal structure. This is the simplest reasonable and therefore also the most plausible explanation for why quantum interaction appears to us as discontinuous in space and time. I'll talk more about this in the next chapter. Suffice for the moment to point out that the randomness of quantum interaction never threatens the causal structure of our world precisely because of its discontinuity in space and time.

If this ontic structure of no-causality-causality represents the starting-point for our world, it must also define the way the principle of noncontradiction initially applies, that is to say, not merely in terms of mutual exclusion, but also joint completion. Although Bohr, like everyone else, seems to have failed to realise this possibility, we can at least borrow from him here and describe this application as defined by the complementarity of no-causality-causality, that is, both their mutual exclusion and joint completion. It is this complementary relationship that can then be taken to constitute the starting-point for our world, including, by implication, the starting-point and initial connection for literally all of our knowledge. Thus, this complementarity even comes to define the initial application of the first law of logic in our world.

> *I'm not really sure I follow all that. You might need to go through it again, I think.*

Sure, let me try and explain it again in more detail. Before I do, though, let me just reiterate that it is this metaphysical implication that almost certainly represents the real importance of our discovery of the quantum mystery. Not only that, but this implication also becomes virtually unavoidable once we come to understand the significance of this discovery on our traditional application of the law of noncontradiction.

Two

The Quantum Mystery and Particle-Wave Duality

Let me explain the basic mystery surrounding quantum interaction. Perhaps the most famous example of this mystery is the particle-wave duality of quantum objects. Do you know what particle-wave duality is?

> *Isn't it how quantum objects act like both waves and particles?*

Yes, but we need to clarify something from the beginning. It's not actually possible to observe directly quantum objects themselves. What we see are quantum events, literally tiny little flashes of light that we can then record and measure. And these little flashes always happen randomly in space and time; in other words, they're always observed as being spatiotemporally discontinuous.

This is in contrast to the accepted understanding we get from theories such as Einstein's general relativity that space-time is entirely continuous. We also refer to this space-time continuum as being governed completely by the law of efficient causality.

These concepts of spatiotemporal continuity and causality are really the cornerstones of modern physics, and they play a big part in why the seemingly contradictory behaviour of quantum objects and quantum discontinuity have proven so difficult to understand.

Remember, though, the key point here is that the relationship between continuity-discontinuity doesn't have to contravene the law of noncontradiction if it is taken to exist before the application of this law. You can even expect the limit of the physical world to appear to us as continuous-discontinuous if the starting-point for our world derives from the emergence of causality from no-causality.

> *I get what you're saying. It does make sense, I suppose, that this relationship could exist before the application of noncontradiction.*

From our location within and as part of the same otherwise continuous, causally governed world such underlying randomness would have to appear to us as discontinuous in space and time. In which case, quantum discontinuity would never contradict the causal structure of our world or the space-time continuum of general relativity.

The spatiotemporal discontinuity of quantum events is really an under-appreciated aspect of quantum interaction. Physicists tend to focus more on the discreteness of this interaction, its lumpiness, so to speak. That's where the label 'quantum' originally came from: discrete packets of energy or quanta. We're still referring to the same quantum events or tiny little flashes of light, and they're still the only thing we observe of quantum interaction. It was the theoretical physicist Max Planck who originally mathematically predicted the existence of these random discrete lumps of quanta in 1900, even before they could be observed.

You've got to love the mathematics! Theoretical physics is pretty much the greatest development of our modern age. The mathematics is not the problem; the trouble is our metaphysical starting-point. When I talk about spatiotemporal discontinuity, I'm referring to the same discrete lumps of energy that physicists talk about. I just prefer to use 'discontinuity' because it tends to place more emphasis on the randomness in space and time of these discrete quantum events. This is really where I believe the answer to the mystery lies, that is to say, the relationship between this discontinuity and the continuous causal structure of our world. Basically, we project what we assume must be caused by real objects based on our observations of these spatiotemporally discontinuous quantum events.

> But, surely, something has to cause the quantum events that we observe. There has to be objects behind them causing the flashes of light that we see. These flashes must be collisions between particles.

That's what we commonly assume, based on our understanding of causality and of space and time. But, remember, all we ever actually see are random flashes. We naturally take these random events to be caused by collisions between real quantum objects interacting with each other in space and time.

Also, a big part of the reason we assume these flashes of light are caused by real objects is because, after recording enough of them, their accumulated traces start to display distinct patterns, like you would expect from either particles or waves. The sort of pattern you get depends on the experiment you do.

Probably the most famous example is the double-slit experiment. Interestingly, this experiment was used a couple of hundred years ago to prove that light must be made up of waves, not particles or corpuscles as Isaac Newton had asserted. We now know that light actually displays the characteristics of both particles and waves, and the double-slit experiment is able to help confirm this fact.

I understand, light is made up of photons.

But remember, nobody ever actually sees these photons. All we ever observe are just discontinuous flashes. We then naturally take these flashes to be caused by the interaction of light photons with other quantum particles.

How the double-slit experiment works is, basically, we fire a beam of quantum particles—or, rather, what we assume to be quantum particles—through two slits in a screen and we get an interference pattern on another screen behind that is consistent with the beam being made up, not of particles, but of waves. The pattern looks like light and dark lines, kind of like a barcode, in which the dark lines are presumably where the two waves cancel each other out.

On the other hand, if we actually observe which slit each individual particle goes through—scientists are even able to do this nowadays—we no longer get the interference pattern on the screen behind. We get what they call a machine-gun pattern, consistent with if we were firing a stream of particles through the two slits. It's like the waves have suddenly disappeared, and it really is particles now going through the slits.

Again, the thing to remember is that what we always observe and record on the screen are just random flashes. Both the interference pattern and machine-gun pattern are always made up of the traces of these discontinuous quantum events, literally thousands of them to make an interference pattern. Even when we observe the individual particles going through the slits, what we really see are just random flashes, either at one slit or the other. We assume there must be real objects behind these events for the simple reason that something must be causing them, and perhaps even more significantly, because of the fact we get distinct patterns forming on the screen that indicate the existence of either waves or particles.

Now, there are two big assumptions at play here that are taken almost entirely for granted.

The first, of course, which we just talked about, is the assumption that the truism of noncontradiction must naturally apply in any such description of quantum objects. But, as I pointed out earlier, this might not be the case for the simple reason that the relationship between spatiotemporal discontinuity-continuity, at the heart of this mystery, could actually come before our application of the law of noncontradiction.

> *I understand. You're saying that the starting-point for our world could be the emergence of causality from no-causality, which would appear to us from within the same world as spatiotemporal continuity-discontinuity. As the starting-point for everything, this relationship would exist before the application of the law of contradiction. In that case, no-causality and*

causality would not simply contradict each other, but would also be jointly completing in the ontic structure of our world.

Now, the second assumption, which really derives from the first, is that quantum objects must somehow lend themselves to a space-time description. In other words, they must ultimately adhere to the law of cause and effect. Remember, causality and spatiotemporal continuity are the cornerstones of modern science. Basically, the metaphysics of modern science presupposes a real spatiotemporal world of moving parts that must be governed by the law of efficient causality and entirely describable through mathematics.

We take this space-time description to be non-negotiable because, if it wasn't, our whole mathematical understanding of the physical world could fall apart.

But this way of thinking really stems from the first assumption, that noncontradiction must naturally apply as an unavoidable truism. We fail to appreciate that the randomness of quantum events doesn't need to be accounted for within the causal structure of our world because it can be understood literally to precede the emergence of any causality. In other words, it could exist in a complementary relationship to the causal structure of our world—and the mathematics we create to describe that causal structure—as a fundamental part of the very ontic fabric of existence itself.

Given that the relationship between these spatiotemporally random flashes, and the otherwise continuous causal structure of our world—represented, for example, by the double-slit experiment and the interference or machine-gun patterns—defines the limit of what we can observe, it is reasonable to suppose that this relationship could also come directly from the ontic starting-point for existence itself in our world. As we have just been talking about, it's quite possible to imagine this structure appearing to us as a complementary relationship of spatiotemporal discontinuity-continuity, deriving from the physical emergence of causality from no-causality. Again, this relationship would have to come before the application of the truism of noncontradiction even as a fundamental law of existence.

So, what I'm saying is that existence itself is defined by this complementary relationship between no-causality-causality.

> Well, I guess that makes sense. If causality emerges from randomness as the starting-point for our world, then it's fair to think that we might still be able to see traces of this randomness. It also makes sense that it would appear to us, from within the same world, as discontinuous in space and time and complementary to our world's causal structure. I suppose, it makes even more sense when you realise that randomness is defined simply as no-causality. First there is randomness, or no-causality, and then causality emerges as the real, physical starting-point for our world.

In our attempts to understand the double-slit experiment, and what we naturally take to be real objects interacting in space and time, we fail to appreciate the complementary nature of this ontic relationship between no-causality-causality. The traces of spatiotemporal discontinuity that we observe in the double-slit experiment, the random flashes of light that we record on the screen or at either of the two slits, do not need to be caused by real objects interacting in space and time. This is a misconception based on our presumption of a causal ontology and ultimately our assumption that the truism of noncontradiction must apply to such objects. Once we set aside these assumptions, the best and most plausible explanation for these random flashes is simply that they represent traces of the spatiotemporal randomness that makes-up part of the real, ontic structure of our world and which exists in a complementary relationship to its causal structure, that is to say, in a mutually exclusive, but also jointly completing relationship.

It's almost certainly a misconception to attribute the interference patterns and the machine-gun patterns recorded on the screen to the interaction of real quantum objects. It's more plausible that they derive instead from the relationship between this spatiotemporal discontinuity-continuity that represents the likely appearance of the real emergence of causality from no-causality and the effective starting-point for our world. Any inability to conceptualise this possibility comes about through our age-old assumption of the need for a causal ontology, based on the application of the truism of noncontradiction to the real structure of our world.

These two assumptions have been almost universally taken for granted for centuries, with some of the most astute and influential minds in Western history presupposing their validity: Aristotle, Hume, Kant, Hegel, Galileo, Newton, Einstein, and Bohr to name just a few of the most prominent. With the discovery of quantum discontinuity, we fail to realise that it is actually these assumptions that have been brought into question. Virtually all of our efforts to understand the double-slit experiment and other such quantum experiments, and what they are most likely telling us about the ontic structure of our world, take for granted the need for a causal, space-time description of quantum objects based on the application of the truism of noncontradiction.

Even our language makes it difficult to avoid this idea of real particles or real waves being the cause of the patterns we see in the double-slit experiment and other such quantum experiments. When scientists and philosophers speak about the quantum realm, they invariably talk about quantum objects, particles or waves, moving around in space and time. For convenience, I just talked like that myself in describing the double-slit experiment.

As I pointed out earlier, our ability to describe the behaviour of quantum objects was the subject of Einstein and Bohr's debate. Einstein believed these quantum objects must somehow be real in space and time, and describable, especially through mathematics, while Bohr thought, maybe, such a space-time description is actually not possible. Bohr, in the end, even came to focus on language itself as perhaps being insufficient for the task of describing quantum objects. But even Bohr never really doubted the actual existence of such objects, and he never questioned, either, the validity of the application of the truism of noncontradiction to them.

So, are you suggesting that we need to create a new language?

No, I don't think that's necessary. Our language has evolved as part of our world, and it makes little sense to me that it would somehow be inadequate for describing that world.

First, we need to recognise the assumptions that are defining our thinking here. Basically, we assume the necessity for a space-time description of quantum objects because we take for granted the application of the truism of noncontradiction to them. In other words, we naturally assume that quantum objects have to be causally governed for the simple reason that, if they were truly random, our whole mathematical description of the physical world could potentially unravel.

Again, this way of thinking hinges entirely on our presupposing that the truism of noncontradiction must apply to quantum objects.

Once we set aside these assumptions about noncontradiction and causality, we can then start to understand the real meaning of quantum interaction. We can start to realise that what we're probably looking at are merely the effects of the relationship between no-causality-causality—to be more technical, spatiotemporal discontinuity-continuity—defining the limit of what we can observe and measure in our world. Our presumptions about quantum objects somehow existing beyond this limit, in some strange quantum realm, are based entirely on our observations of this relationship, and of course, also from our natural assumptions about the need for causality and noncontradiction to be accounted for beyond this physical boundary.

Three

The Application of Noncontradiction is No Longer an a Priori Certainty

The possibility that the starting-point for our world could be the emergence of causality from no-causality, and thus effectively the complementary relationship between spatiotemporal continuity-discontinuity from our location within and as part of the same world, means that the traditional application of the truism of noncontradiction can no longer be taken simply as an a priori certainty in our world.

This really is the key point to get your head around. Although the emergence of causality from no-causality as the physical starting-point for our world may amount only to a theory, the mere possibility of this scenario, and the fact that such a relationship would have to precede the initial application of the law of noncontradiction, renders the a priori certainty of this law open to question. Both our traditional philosophical and modern scientific understanding of the metaphysical structure of our world are predicated on the supposed a priori certainty of the application of the law of noncontradiction as a straightforward truism. For this reason, the mere possibility of this scenario, made plausible by the discovery of quantum discontinuity at the measurable limit of our world, not only brings the possibility of a priori knowledge about our world into question, but causes our traditional foundation of knowledge itself to become problematic.

Again, a truism is always a truism. A contradiction is a contradiction. But, for the truism of noncontradiction to serve as a law in our world, it first needs to be applied as such. If the starting-point for absolutely everything is the real emergence of causality from no-causality, then this relationship would have to come before any possible application of the law of noncontradiction. Because we experience this physical relationship from within and as part of the same world, it appears to us as spatiotemporal continuity-discontinuity.

It's not just that this relationship existed before we did, but it would have literally had to come before everything, including any possibility of applying any law in our world.

This is how such physical randomness becomes possible in our otherwise causally governed world.

Quantum randomness never contradicts the law of efficient causality because it precedes it. It is this complementary relationship between continuity-discontinuity, that is to say, its mutual exclusion and joint completion, that can be reasonably taken to constitute the real, ontic starting-point for our world.

It's significant, as Aristotle pointed out, that ancient philosophers interpreted the world as starting from this relationship of continuity-discontinuity. They interpreted our world this way because this is how it appears.

It has really only been because of our efforts to accommodate the truism of noncontradiction as the first law of logic that has clouded our ability to appreciate this ontic starting-point, not to mention, its metaphysical implications.

Once we lose our certainty about the application of the truism of noncontradiction to our world, the a priori application of this truism to metaphysics, to logic and even to mathematics becomes uncertain. 'A priori', that is, 'prior' or earlier knowledge from the Latin, derives from logical laws, starting traditionally with the law of noncontradiction. It's knowledge that is presupposed not to require any experience, based on so-called first principles that are self-evident and unavoidable. This has been the methodological approach of metaphysics since Aristotle, and it served to connect, for example, the logical theories of Kant and Hegel to our world.

However, and again, this really is the point, once this initial connection to our world of the law of noncontradiction becomes uncertain, then a priori knowledge about our world must also become uncertain. The best we can possibly hope for under such circumstances is to weigh up the evidence based on our experience of the world and Occam's razor.

> That's what we're doing here.

You really need to get your head around this realisation!

> I'm not sure I entirely understand. So, are you saying that quantum mechanics is all wrong?

Of course not. As I said before, what we're getting at actually has very little to do with quantum mechanics. Look, quantum mechanics has provided some of the most successful mathematical theories in the history of physics. Again, by quantum mechanics I'm just referring to the mathematics used to describe and predict the effects of quantum interaction in our physical world. It's sufficiently connected to our world by the fact it describes our observations and what we can physically measure. It successfully meets Occam's razor, if you like.

Even though pure mathematics is a form of deductive reasoning, it's not reliant on its apriorism when used to describe the physical world. The significance of this point is not always appreciated in contemporary thinking. Even such seminal figures as Descartes, Galileo and Kant presupposed this absolute function of mathematics in our descriptions of the physical world. Galileo, for example, believed mathematics to be the language of God, and that God made the physical world as an immutable mathematical system that could be known absolutely through the divine language of mathematics. One of the reasons Galileo got into trouble with the Catholic church was that he argued for the authority of mathematics to be able to interpret the holy scriptures as the Word of God.[1]

Newton, on the other hand, never applied mathematics in this way. Newton certainly believed in the power of mathematics to describe the fundamental workings of the physical world, but he never took the connection of mathematics to our world to be in any way a priori, or the ability of mathematics to describe the physical world to be dependent on its apriorism. For Newton, such mathematical descriptions had to be always referred back to our experience of the world for their validation. This is something that is often not clearly understood in our modern thinking because we take the metaphysical world to be entirely logical. Newton was astute enough to realise that even our most seemingly self-evident assumptions could be open to metaphysical question.[2]

The correspondence of mathematical descriptions to the physical world needs to be verified through observation and experimentation. This is why we refer to Einstein's theory of relativity, for example, as a 'theory': it remains falsifiable if better evidence, or a better explanation can be found.

This is why mathematical theories like you find in quantum mechanics and their relevance to our world do need to be verified through experimentation. We understand this as part of the scientific method. It is not simply its apriorism that connects quantum mechanics to our world, even though its power as a deductive method certainly contributes to our accepting its descriptive and predictive accuracy. What we observe in quantum experiments is real, it's measurable, and of course, it's describable using mathematics.

It's when physicists try to reach beyond the limit of what is measurable that they start to get themselves into trouble. This is because they then move from the measurable world, to not just a presupposed quantum realm, but to the ontic structure and potential starting-point for existence itself. Certainly, physicists continue to adhere to the scientific method and endeavour to accord their calculations to our observations, but they also continue to presuppose the natural validity of the truism of noncontradiction. As Aristotle originally asserted, this is what traditionally connected knowledge and particularly a priori knowledge to our world.

By doing this, physicists are also presupposing certain requirements that need to be maintained with regard to this starting-point for our world; most significantly, it must continue to adhere to a causal ontology, even in spite of our observations of quantum discontinuity and duality. The mathematics then effectively becomes more dependent on its presupposed a priori aspect rather than observations to justify itself and to maintain its connection to our world. In short, mathematics starts to overshoot its usefulness.

This is then compounded by the fact that we presuppose quantum objects to be subject to a space-time description, in other words, again, we assume a causal ontology ultimately based on the application of the truism of noncontradiction to this supposed quantum realm. This is really where the difficulty lies in attempting to formulate a mathematical theory of everything: it's in trying to reconcile the quantum discontinuity we observe in quantum experiments with the presumption of a causal ontology, based on what we naturally presuppose to be the unavoidable truism of noncontradiction.

So, on the one hand, we take for granted the application of the truism of noncontradiction to our world, and on the other, we naturally assume this truism to be the a priori starting-point for any knowledge about our world. This is all justified by the supposed certainty of the application of the truism of noncontradiction to our world. Can you see where the problem is when we come to try and understand the starting-point for it all?

> *Well, I guess, by taking for granted the truism of non-contradiction, we're already effectively pre-defining the starting-point we're trying to figure out.*

These days, we take the validity of the application of mathematics to the ontic structure of our world very much for granted. As I said, this question was at the crux of the debate between Einstein and Bohr, with Einstein basically following Hegel's lead and arguing that this ontic structure should be entirely describable using mathematics and logic. Like Kant, Hegel followed the Western metaphysical tradition of starting with the a priori application of the truism of noncontradiction to our world. Hegel's dialectic was part of his solution to this age-old dilemma. We'll talk more about Hegel in Chapter Seven.

But, again, what is the empowering mechanism behind this assumption?

> As you said, it's noncontradiction as the a priori first law of logic?

In Bohr's case, in saying that mathematics may not be able to describe the quantum realm, he wasn't arguing, like we are here, that there may be a problem with our application of the law of noncontradiction. Bohr was following Kant's lead and asserting that we may not have cognitive access to this quantum or noumenal realm. Kant originally came to this conclusion basically because such a noumenal realm would have to be the ultimate source of contradiction in our world and beyond the limits of our cognitive ability to comprehend. Similarly, Bohr argued that it might not actually be possible to comprehend the quantum realm due to the simple possibility that there may be a natural limitation to our thinking.

> I guess there is a limit to what we can see from within the same world. As you're saying, we experience this emergence of causality as the relationship between spatiotemporal discontinuity and continuity.

And Kant basically understood this. He didn't know about quantum interaction, of course, and there is no reason to think he suspected the emergence of causality from randomness as being behind it all.

In Kant's time, they had a similar conflicting explanation for the ontic structure of our world. The two intellectual giants of the time were Newton and Gottfried Leibniz. Although these guys had both been dead for the better part of a century, their ideas were still dominating Europe at the time of Kant. To keep it simple, Newton represented the new, modern scientific way of doing things and Leibniz the old metaphysical way of sitting back in his armchair, thinking deeply about the world. Newton advocated a causal ontology and the modern scientific notion of a continuous causal structure to the world, while in contrast, Leibniz reasoned that the world had to be somehow fundamentally discontinuous.[3]

Kant concluded that the only way to reconcile these competing perspectives was to separate the phenomena from the noumena.

Where Kant came unstuck was in presupposing that we could somehow have a priori knowledge of the real world. He based this presupposition on the initial application of noncontradiction as an a priori truism. Indeed, it's this aspiration for a priori knowledge, based on the a priori application of the truism of noncontradiction, that has really been the downfall of Western metaphysics since Aristotle, and it still dominates both analytic and continental philosophy to this day. Put bluntly, I believe it's basically the reason why the study of metaphysics has always struggled for credibility and really the root of why contemporary Western philosophy is perceived to lack relevance beyond academia. Unfortunately, modern physics tends to fall into the same trap when it comes to trying to figure out the ultimate theory of everything.

The law of noncontradiction is supposed to be an a priori certainty. It's supposed to be an absolute law requiring no evidence. According to Aristotle, and perhaps for obvious reasons, the impossibility of something contradicting itself is the most certain thing we can know. That's why Aristotle identified it originally as the first law of logic. It's this natural apriorism and absolute certainty that has always assured us of the logical validity of noncontradiction and endowed it with the power to serve as the first step in a priori methods of analyses.

Again, however, it's not actually the truism itself of noncontradiction that validates it as the first law of logic in our world, but the necessary application of this law specifically to our world. This is the mix-up we've been making since Aristotle.

What the first law of logic is supposed to do is connect such a priori methods to our world and ensure their validity specifically to our world. This is an aspect of the law of noncontradiction that has always been taken for granted in Western thinking.

The strength of the scientific method, on the other hand, is its reliance on observation and experimentation to verify the mathematical theories of theoretical physics. This works well with quantum mechanics when we're simply trying to describe and predict the effects of quantum interaction in our world. It becomes less clear, however, when we start to apply mathematics beyond what we can actually observe, and when the thing we are trying to account for is the very ontic structure and potential starting-point of our world. This is especially the case when the evidence we are attempting to mathematically describe appears to contain a contradiction.

I'm talking about attempts to come up with the so-called mathematical theory of everything: strings and super strings, membranes, many worlds and many dimensions, observer created realities, etc. Remember, this is the problem that Einstein applied his immense brain power to, unsuccessfully, for the better part of his life.

Such a priori theories purport to account for the physical evidence, particle-wave duality, for example, and they may even do that adequately, but what is really empowering them and presumably validating them as logically relevant to our world, and particularly to its ontic structure, is the initial application of the truism of noncontradiction as an a priori law. Without this validation, literally the initial step connecting the mathematics to our real world, they no longer have the power as a priori theories and end up needing to be judged solely on how well, and especially how economically, they are able to account for the evidence.

In other words, they become very much subject to Occam's razor! The best explanation is the one that relies on the fewest assumptions.

> And, as you're saying, the emergence of causality from no-causality is the simplest plausible explanation for the existence of quantum discontinuity. But does this explanation rely on the least assumptions?

Well, it starts out by trying not to assume anything, and to give the simplest possible explanation based on the best evidence, the best evidence being the relationship between spatiotemporal continuity-discontinuity that is acknowledged as defining the measurable limit of our world.

Look at it this way, as I said before, by relying on the truism of noncontradiction we're already assuming certain things about the ontic structure of our world, that it must be causally governed, for example, and that it has to be subject to a space-time description.

In other words, we're already pre-defining this ontic structure and, by implication, the starting-point itself for our world.

Perhaps not surprisingly, it's this natural presumption that the truism of noncontradiction is an absolute certainty in our world that has created so much of a headache for philosophers throughout the history of Western metaphysics. It was this apriorism, as I said, that really drove Kant to conclude that there must be something more beyond the phenomena we can experience, what he referred to as a noumenal realm, where, basically, contradiction must originate from and can be accounted for. This was also what Hegel then set out to overcome in his systematic theorising.

The point is, how can this law remain an absolute certainty when there is a real possibility that the relationship between no-causality-causality comes before its application? The key is to realise that we've always mixed-up the application of this law with the idea of it as a simple truism. How can this law hold up as an *absolute certainty* when the ontic structure of existence itself could actually be defined by a complementary relationship? Remember again, we're not questioning the idea of noncontradiction as a truism, but only its necessary application as a real law in our world.

A law can only serve as a law after it has been applied as such. It's this uncertainty about the application of noncontradiction as a fundamental ontological law that represents the real significance of the discovery of quantum randomness.

Basically, we're talking about the historical foundation of Western thinking itself. Philosophers may sometimes question the role of this truism as the first law of logic, but they never seriously question the validity of the truism itself as an absolute certainty or as a basic law of existence. To do so is considered logical heresy.

In questioning the role of noncontradiction as the first law of logic, it's invariably logic itself that is applied, that is to say, an a priori method that naturally presupposes the initial application of noncontradiction as a logical truism, quantum logic, for example. Maybe that's why this law has managed to hold up, even in the face of the discovery of quantum discontinuity. That and the fact we've always mixed-up the idea of noncontradiction as a simple truism, and the application of it as a real law in our world.

It seems less dangerous, I guess, in trying to solve the quantum mystery, to talk about the possibility of many different worlds or an observer created reality than to risk the very foundation of our thinking itself.

So, does everything fall apart?

Of course not. We're talking about adding something to our current understanding of the world: randomness exists in a complementary relationship to causality; no-causality-causality are not simply mutually exclusive, but also jointly completing in our world.

This is a very empowering realisation when you think about it. It doesn't threaten anything, and certainly not the causal structure of our world.

It's really how we experience the world, anyway! And, it does open up the possibility of solving some of our most enduring mysteries, starting with the enigma of quantum discontinuity.

It's fair to say that the study of Western metaphysics has always pretty much failed to live up to expectations, maybe for the simple reason that its starting-point has always been wrong. It has always aspired to obtain a priori knowledge of our world, starting from the absolute certainty of the truism of noncontradiction.

Kant, for example, set out to turn Western metaphysics around and to put it on a scientific footing comparable to the physics of Galileo and Newton. It's perhaps not surprising that Kant started out as a mathematician. And, of course, his aspiration was to obtain a priori knowledge of the real world starting from the absolute certainty of the truism of noncontradiction. Significantly, when faced with the need to account for what he concluded to be a necessary element of spontaneity or randomness actually built into our thinking itself, the best Kant could come up with was something he ambiguously referred to as 'so-called mother wit'.[4]

Four

Quantum Discontinuity is an Ontological Problem

As I said at the beginning, the reason we struggle to understand the meaning of quantum duality is because we mistakenly assume it to be a problem for mathematics. By taking the application of the law of noncontradiction for granted, we already effectively pre-define the ontic starting-point for our world and naturally presuppose its ontic structure to be causally governed.

There is no question that the physical world is causally governed. As I'm arguing, the most plausible explanation for the starting-point for our world is the emergence of causality from no-causality (i.e., randomness). From within and as part of the same world, it makes sense that our world looks entirely causal and continuous in space and time, and that any inherent randomness in the ontic structure of our world would appear to us as spatiotemporally discontinuous. The point here is that causality emerges originally from no-causality, in other words, a state of simple randomness.

THE REAL STARTING-POINT FOR OUR WORLD

It needs to come from somewhere, I suppose.

When you're able to look closely enough, via quantum experiments, for example, you find the limit of our physical world to be constituted by traces of no-causality, that is to say, discrete, discontinuous events in space and time. And, you realise, not only is there observable traces of spatiotemporal randomness, but that such inherent randomness is a fundamental and necessary part of the physical make-up of our world. In other words, it doesn't just appear as somehow contradictory and mutually exclusive to the causal structure of our world, but it also actually has a jointly completing role to play in it, as a complementary and fundamental aspect of our world's ontic make-up.

You come to realise that this relationship between no-causality-causality most likely constitutes the real starting-point of our world and existence itself.

> *It does make sense. Randomness appears to us as complementary because the ontic starting-point for our world is made up of the relationship between causality and no-causality. So, does this mean there's no quantum realm?*

Well, not as we assume. What we commonly imagine to be quantum objects, some sort of hybrid particle-wave, for example, are most likely just the product of this underlying relationship between no-causality-causality.

Then you start to realise that the entire atomic structure of our world is composed of this random-causal relationship. Bohr is famous for being the first to work out the real structure of the atom, and he did this by following Planck's lead and quantising it.[1] This just meant that Bohr allowed for this discreteness and spatiotemporal discontinuity; in other words, he allowed for the existence of physical randomness in space and time.

When you think about it, that's most likely all that fundamental particles are! Electrons, photons, etc. They *are* spatiotemporal randomness (i.e., discontinuity, discreteness), or traces of no-causality in the ontic fabric of our world! At least, as that randomness interacts with and is contained by the causal structure of our world. Planck and Bohr just incorporated this random aspect of the real world into their calculations—even if their ingrained metaphysical assumptions wouldn't allow them ever to understand it that way.

Unfortunately, to think of the discreteness, the discontinuity, as inherent traces of no-causality runs in the face of not just classical physics and the idea of causality as an absolute law governing our world, but also our traditional understanding of metaphysics. Instead of recognising this discreteness and discontinuity for what it almost certainly is, we've continued to try and fit it into our age-old metaphysical assumptions about the world.

> *Okay, but I still don't understand why nobody else has picked up on this. The way you've explained it seems pretty straightforward. If the emergence of causality from no-causality is the starting-point for everything, then this relationship would have to come before the application of noncontradiction. That makes sense, and, if you're correct, it seems to account for the existence of quantum discontinuity. So why is it such a mystery?*

Well, that's a good question. For one thing, as I said, we naturally approach the mystery as a problem for mathematics, when, really, the ontic starting-point for our world is a question for ontology. It's metaphysical not physical.

As I also pointed out at the beginning, Western philosophy has pre-occupied itself mostly with epistemology for the past several hundred years. Remember, epistemology is all about the study of how we know the real world, and it's aspiration has always been to attain a priori knowledge of our real world. The metaphysical starting-point for this has always been the logical truism of noncontradiction, with its supposed strength lying in the fact that, as a logical truism, such a starting-point doesn't require any reference back to our experience of the world.

As I said, this question of how we know the real world was at the heart of Einstein and Bohr's debate. In fact, the vast majority of metaphysical studies of quantum interaction naturally take an epistemological approach: they mostly enquire into how we can know quantum objects, assuming, in the process, that we can't figure out the real structure of the quantum realm until we're able to answer this epistemological question. When they do consider the ontic structure of the quantum realm, they commonly take for granted the application of the truism of noncontradiction to it and mostly attempt to figure it out mathematically.

Our epistemology presupposes the application of the truism of noncontradiction. Epistemology applies an a priori method based on the assumed certainty of first principles, with the most secure of all being the truism of noncontradiction.[2] That's been the starting-point for Western metaphysics for the past two and a half thousand years! It aims to work out how we can know the ontic structure of our world, as the source of first principles such as noncontradiction. We not only take its application for granted, but by doing so, we're already presupposing certain stuff about this ontic structure: that it can't be contradictory or random, and it must somehow be subject to a space-time description.

Descartes really set the tone for modern thinking and metaphysics with his question about what we can know with certainty.

> *I think, therefore I am!*

Descartes arrived at our own thinking as the one thing we could know to exist with certainty.

It's worth noting that Descartes came to this conclusion in a book he titled *Meditations on First Philosophy* (1640). His method was explicitly epistemological, and his aim was to arrive at an ultimate foundation for knowledge. Galileo had just been prosecuted by the Roman Inquisition for teaching ideas contrary to the Catholic Church. The dominant philosophy of the Church was Aristotelean empiricism, which held essentially that our experience of the world, our senses, provided the most certain form of knowledge. Descartes argued a rationalist approach, asserting that our senses may be deceived and that the only truly indubitable form of knowledge derives from our own rational thinking.

In applying his epistemological method, Descartes already assumed the validity of the truism of noncontradiction. For Descartes, the correct foundation for knowledge was indubitability, knowledge that was patently evident, and he presupposed such knowledge to be possible because he naturally took for granted the straightforward application of the truism of noncontradiction to our world. This is implicit throughout his *Meditations*. Descartes had no reason to doubt its application in our world. His point was that he didn't want to rely on experience; as he surmised, we may be getting deceived by an evil demon.

The history of modern Western philosophy has been driven by this epistemological aspiration for indubitable knowledge of our world, whether empiricist or rationalist. The rise of modern physics and its successful reliance on mathematics has fuelled this aspiration and driven contemporary philosophy to try and emulate this success through the application of logic to our metaphysical world.

The same with the rationalist approaches of Kant and Hegel. Their ultimate goal was to figure out what we can know about the world with certainty, and they always took for granted the application of the truism of noncontradiction as their initial starting-point.

It's because of this enigmatic starting-point that professional philosophers over the years have homed in more and more on our own cognition as holding the key to this problem. With their analyses presumably tethered sufficiently to our world by the truism of noncontradiction, they've come to be satisfied that it's our own thinking that must hold the key to unlocking the metaphysical world.

> *But isn't there a difference between analytic philosophy and continental philosophy?*

Yes, certainly in the detail and how they go about their analyses. Both analytic and continental philosophy are very similar, though, in the common way they both ultimately aspire to a priori knowledge of the real world, they both take for granted the natural application of the truism of noncontradiction, and they both assume the key to the ontic structure of our world lies in our own cognition.

I think it's fair to say that Professional philosophy has come to be dominated these days more by the Anglo-American analytic way of doing things. At least, this is the case in the English-speaking world. The majority of mainstream academic journals in English won't even consider philosophical essays that are not grounded in this analytic method. It's all about small, incremental steps, building up our a priori knowledge of the real world piece by piece through rigorous logical analyses. It's perhaps not surprising that this method has its origins in mathematics, again, really trying to emulate the success of modern physics.

Gottlob Frege, commonly recognised as the founding father of analytic philosophy, was a professor of mathematics. Frege introduced language into the mix. I mean, he wasn't the only one to focus on language, but what set him above the rest was his genius as a logician. We're talking about the late nineteenth century, just prior to the discovery of quantum interaction. Frege is recognised as the greatest logician since Aristotle. He changed the game, basically, with his mathematical approach to language.

As logic deals in the business of working out the laws of thinking, philosophers have come to satisfy themselves that these logical laws must also govern the metaphysical world, in the same way that the physical world has been shown to follow the laws of mathematics. And, as the laws of logic ultimately come to us in the form of language, it makes sense that philosophers would focus in on the link between language, thinking and the ontic structure of the world.

After all, it does make sense that the rules of thinking and the rules that govern the real world would tend to correspond. We have evolved in the same world and are a part of it. I have no problem with that. Again, the only issue here really is the mixing-up of the idea of noncontradiction as a mere truism and its application as a fundamental ontological law in our world. Once we recognise this as being the real import of the discovery of quantum discontinuity, the starting-point for everything changes from simple mutual exclusion to both mutual exclusion and joint completion.

We also come to understand that this change becomes unavoidable after you realise the implications of this discovery on the a priori certainty of this age-old first law of logic.

It's no coincidence that language happens to be governed by a similar complementary starting-point: the linguistic structure (langue) and the speech act or event (Parole). This structure was first identified by the French linguist Ferdinand de Saussure in the generation following Frege. Saussure's linguistics is an interesting side-street to go down in trying to understand some of the implications of what I'm talking about, but one I don't intend to follow here.[3]

The aspiration in modern philosophy has been to attain a priori knowledge of the real world and the path to it began as an epistemological enquiry into how we can obtain such a priori knowledge. This was the legacy of Descartes, and it was exactly how Kant and Hegel, for example, went about their philosophising.

There is no need for me to go much further into analytic philosophy. We can think of it as the end product of this epistemological approach to metaphysics in Anglo-American philosophy. Suffice that such an approach will always struggle to arrive at what we're talking about because it starts out by assuming the application of the truism of noncontradiction to be an a priori certainty. By doing so, it already effectively pre-defines its own starting-point. It really needs to begin by setting aside this starting-point as a first principle, but it can't do that because it aspires to be a priori.

The analytic method is certainly capable of questioning the role of the truism of noncontradiction as the first law of logic, but it never seriously doubts the application of this truism itself to our world. To do so it considers logical heresy. It takes the truism of noncontradiction as already established as an a priori first principle, brooking no argument and requiring no further evidence.

By approaching it this way, the analytic method conflates the application of noncontradiction as a fundamental law in our world with the idea of it as a logical truism.

What we're really talking about here is first philosophy. Remember, I mentioned earlier how Aristotle originally referred to his metaphysics as 'first philosophy'. Perhaps understandably, this title got left behind and came to be replaced with the term 'metaphysics' in the sense that it's the field of study that is supposed to go beyond or transcend the limits of the physical world. Strictly speaking, it acquired the name metaphysics as referring to the text that came after *Physics* in Aristotle's works. It doesn't really matter, suffice that the idea of 'first philosophy', as our initial theorising about the ontic structure of our world, got lost along the way.[4]

You could say that Aristotle was too convincing in his original argument that the truism of noncontradiction must serve as the first law of logic. The initial job of metaphysics, that is to say, to actually figure out the starting-point and ontic connection for any knowledge to our world, was a done deal before metaphysics, as a philosophical discipline, even got off the ground. Western metaphysics has struggled ever since.

This question about the starting-point for knowledge is not an epistemological problem, but an ontological one: we've got to decide on our starting-point for knowledge about our world before we can figure out how we are able to know anything about the world, especially its real, ontic structure. We've taken this starting-point for granted since Aristotle because we've always presupposed the application of the truism of noncontradiction to our world. It's not even a 'chicken and egg' scenario. If you're not sure, or at least satisfied about the starting-point and how knowledge actually connects to our world, how are you supposed to know anything about that world?

You certainly can't do it a priori or with indubitability. It's the self-evident certainty of the truism of noncontradiction that has always ensured the connection of knowledge to our world.

Following Descartes, Western metaphysics focused on indubitability as the foundation for knowledge, and it was able to do this because it had already taken the application of noncontradiction to our world for granted.

So we can't know anything?

Of course we can. Knowing things a priori or indubitably might be a bit much, but we can certainly claim to know things based on our experience and the best evidence. That's all that modern science aspires to, anyway. It really boils down to a question of what is reasonable based on the best evidence, with the understanding that any knowledge, potentially even any law or principle, could be subject to falsifiability?

Following the discovery of quantum discontinuity, it may be time for philosophy to accept that the notion of first principles, and its age-old aspiration to a priori knowledge of our world is archaic. It's this misguided desire for a priori knowledge of the real world that has led metaphysics astray since Aristotle and certainly following Descartes.

It's our mixing-up of noncontradiction as a simple truism and its real application to our world that has mistakenly self-assured us of the possibility of such a priori knowledge.

Five

The Critical Ontology of Nicolai Hartmann

My approach to the quantum mystery has been inspired by the German ontologist Nicolai Hartmann. He was around in the first half of last century, the same time as Bohr and Einstein. Hartmann is not so well known these days, although more of his work is starting to be translated to English. Along with Martin Heidegger, he's probably best remembered for helping revive ontology and for putting the study of Being and existence back on the metaphysical agenda.

> *I've heard of Heidegger. Didn't he write a book about Being and Time?*

Hartmann and Heidegger were colleagues for a short while at Marburg University in Germany during the early nineteen-twenties. Hartmann was Heidegger's supervisor at Marburg. I don't think they liked each other very much. Heidegger was a very radical sort of character, and super smart, of course. Hartmann came across as much more conservative, on the other hand. Even his ideas about ontology might, at first glance, seem conservative. But really, they're implications were arguably more radical than Heidegger's in the metaphysical scheme of things. They both developed their ontologies around the same time, basically as a response to the epistemological approach of neo-Kantianism that was prevalent in Germany and particularly Marburg at the time.

As you might expect, their two approaches to the question of Being were very different. I'd argue that Heidegger's approach to ontology, at least around the period of his famous book *Being and Time* (1927), although certainly radical, was still very much in the epistemological tradition of modern Western philosophy. His phenomenological approach to Being was supposed to provide the answer to this epistemological question about how we can know the real structure of the world, just as logic held a similar key for analytic philosophy.

Heidegger and Hartmann were working within what has since come to be known as continental philosophy. Hartmann's work has also been linked historically to phenomenology, although I tend to suspect this is more a result of the popularity of phenomenology at the time than Hartmann's strict adherence to a phenomenological method.[1] Both phenomenology and analytic philosophy work on first principles, they aspire to a priori knowledge, and they both take for granted the application of the truism of noncontradiction to our world. The main difference between the two is that the former sees the answer lying in our a priori experience of the world and the latter in the application of logic.

Heidegger's approach to the question of Being was grounded in this traditional aspiration for a priori knowledge based on first principles while Hartmann accepted that such knowledge about the ontic structure of our world may be impossible. This makes all the difference once our application of the truism of noncontradiction is brought into question.

It's interesting that the founder of phenomenology, Edmund Husserl, arrived at his method of analysis following a crisis in his thinking, basically deriving from this same problem that has plagued metaphysics since Aristotle, of how to account for the existence of contradiction.[2] This problem of contradiction was Kant's starting-point, and like Kant, Husserl finally decided on a so-called transcendental approach that effectively put consciousness before the real world. It's perhaps little surprise that Husserl also started out as a mathematician, his original PhD being about mathematics, and his first book being the *Philosophy of Arithmetic* (1891). Interestingly, Frege criticised this text for what he claimed to be its psychologism: according to Frege, it made the mind itself the source of first principles, such as the truism of noncontradiction.

> So, what's the difference between Husserl and Heidegger's phenomenologies?

Put simply, Husserl focused on our cognitive or mental experience of the world and Heidegger went the next step and considered our pre-cognitive, or everyday lived experience, what he referred to as our Being-in-the-world. For Heidegger, cognition was only a part of this bigger lived experience and needed to be understood within this wider context. Some commentators have argued that Husserl really thought about this idea first anyway and Heidegger just copied it from him.[3]

All brilliant and wonderful stuff, but we're diverging from our main point here. I want to focus on Hartmann's ontology. Unfortunately, even in his day, Hartmann came to be overshadowed by Heidegger and Husserl. You have to realise that these guys were professional rivals; they never missed an opportunity to have a dig at each other and to criticise each other's work.

Heidegger comes across as the more enigmatic personality, and perhaps that explains part of his appeal. This enduring popularity was in spite of Heidegger being heavily criticised for becoming a Nazi in the 1930s and even apparently denouncing Husserl, his mentor and patron, as a Jew. It's sadly fortunate, I guess, that Husserl had retired and died of old age by 1938, before the worst of the holocaust.

Hartmann, on the other hand, rejected the Nazi ideology and, as the story goes, was one of the few professors who refused to begin his lectures with the Nazi salute. An intelligence officer on the Eastern Front for most of WWI and a highly respected Prussian academic by the time of Hitler, I guess Hartmann was pretty much untouchable, even by the Nazis. Heidegger managed to avoid serving for most of WWI, due to health reasons, but worked in the last year as a meteorologist on the Western Front. As a testament to Hartmann's character, and his unflappability, Hartmann apparently spent the final stages of WWII in Berlin writing his last book, *Aesthetics*, a work about beauty, while his world literally crashed down around him.

At first glance, Hartmann's critical ontology may come across as mundane compared to Heidegger. Like Aristotle, Hartmann focused on the categories or the various parts of Being, with paired categories like discreteness-continuity being identified by Hartmann as the most fundamental.

Hartmann's works on ontology are a worthwhile read if you're interested in this stuff, and they're written in much clearer prose than Heidegger's. As an introduction, I'd recommend *New Ways of Ontology* (1942), in which Hartmann gives a general overview of his ontology. We don't really need to go into the detail. The crucial point is that, unlike Heidegger and Husserl, Kant and Hegel, analytic philosophy and virtually the rest of Western metaphysics, Hartmann's ontology did not ascribe to an a priori method.

The key for Hartmann, when it came to trying to understand the question of Being, was to not take anything for granted, including first principles such as the application of the truism of noncontradiction. For Hartmann, the real structure of our world simply 'is what it is', and we should not assume anything in advance about it, even whether or not it adhered to a law as seemingly obvious as the truism of noncontradiction. I think this aspect of Hartmann's ontology tends to be under-appreciated by commentators on his work, and they don't always give it the significance it deserves.

This way of approaching the problem of existence makes all the difference. For Hartmann, to attain a priori knowledge of Being was almost certainly impossible. The best we could hope for when it came to the ontic structure of our world was to theorise based on a close study of the evidence. That's what we're trying to do here. The evidence for Hartmann came first from the phenomena we experience, but also with regard to the scientific and philosophical history of ideas about that experience.

The first step in Hartmann's method was to identify hidden assumptions in our history of ideas.

> *And that's what you've done. The application of the truism of noncontradiction is an assumption that we take for granted in all of our thinking.*

This assumption makes all the difference when it comes to trying to work out the meaning of quantum interaction. As I said before, it underpins the belief in a causal ontology.

The significance of these two assumptions about the truism of noncontradiction and a causal ontology cannot be over-stated. Almost everyone apart from Hartmann has taken them for granted when it comes to understanding the structure of our world.

Hartmann referred to his critical ontology as first philosophy, or at least, he thought that's what best described it, even though he doubted he could get away with calling it that.

> *Did Hartmann talk about quantum interaction like you are here?*

Not in the same way. Hartmann's analysis of quantum physics is quite complex, and there is no need for us to go into it here.

My main criticism of Hartmann's ontology is that he failed to grasp the significance of the discovery of quantum discontinuity on our understanding of the ontic starting-point for our world. The key point we're arguing here. I think Hartmann tended to get too bogged down in the detail. The guy was an encyclopaedia!

According to Hartmann, because we couldn't know the real structure of our world with certainty, we needed to exhaustively map the categories of Being before we could draw conclusions about the ontic make up of our world. In other words, we couldn't arrive at the ultimate starting-point for our world until we knew everything there was to know about its ontic structure. And, of course, in the end, that's impossible. You start to understand why some critics of Hartmann's ontology have dismissed it as just a kind of cataloguing of Being.[4]

As I'm arguing, the best evidence for understanding the ontic structure of our world and its starting-point is the relationship between the spatiotemporal discontinuity of quantum interaction and the continuous causal structure of our world.

> *I understand, because that defines the limit of the physical world.*

Exactly! It's not just because we lack the technology to see further. Quantum discontinuity represents a commonly acknowledged limit to what we will ever be able to measure in our world.

It makes sense when you think of this spatiotemporal discontinuity merely in terms of the fundamental randomness that it most likely represents. The idea of no-causality only really has meaning in its relation to the causal structure of our world. There can literally be no meaning without causality.

So, you can say, the limit of our world is marked by quantum discontinuity, but, really, it's defined by the relationship between spatiotemporal discontinuity-continuity. It's impossible for there to be anything, or for us to even form any cogent thoughts about anything, without causality to underpin it. Quantum discontinuity as spatiotemporal randomness or no-causality can only really have meaning with reference to the causal structure of our world.

Again, it's significant that ancient philosophers understood this relationship of discontinuity-continuity as defining the starting-point for our world. It's only after Aristotle applied the truism of noncontradiction as the first law of logic that we started to lose sight of the significance of this real starting-point.

According to Hartmann's critical method of ontology, the first thing we need to do is to identify the hidden assumptions getting in the way of our analyses. We've done that by identifying the age-old mix-up between the truism of noncontradiction and its application to our world as a fundamental ontological law. Once we bring into question the law of noncontradiction, we can then start to question the modern assumption of a causal ontology.

This also brings into question the very legitimacy of a priori methods of analyses for studying the ontic starting-point for our world. This includes mathematics and logic, and even traditional epistemology and phenomenology. Philosophers since Aristotle have commonly approached metaphysics and the question of Being as an a priori problem. This is still effectively how theoretical physics and mainstream philosophy approach the mystery of quantum interaction.

By doing this, they're already pre-defining the ontic structure of our world that they're trying to figure out.

Once we've identified and suspended the hidden assumptions in our thinking, the next step in Hartmann's ontology is to study the evidence, including both the phenomena and the history of ideas on the given subject, without the assumptions clouding our understanding. The phenomena inevitably draw us to the relationship between quantum discontinuity and the continuous causal structure of the physical world as defining the limit of what we can experience in our world. Because this relationship defines the measurable limit of the physical world, it's reasonable to accept it as also representing the appearance of the ontic starting-point for our world.

This becomes even more reasonable when you remember that a priori analyses have become problematic, particularly with regard to the ontic make-up of our world. In other words, when it comes to understanding our world beyond what can be observed and physically measured, the best you can hope for is to take the phenomena as they appear to us, without presupposing anything, including the truism of noncontradiction or causality, and then apply Occam's razor.

The final step is to draw the best theory based on the evidence and without the hidden assumptions.

> *And, of course, you've done that by concluding that the simplest and most plausible explanation for the existence of quantum discontinuity is the emergence of causality from no-causality as the starting-point for our world.*

This idea of causality emerging from randomness is not new. Perhaps the best contemporary advocate for this explanation of quantum interaction is Arkady Plotnitsky, an American mathematician and professor of literature and philosophy. Plotnitsky provides perhaps the most thorough contemporary analysis of the quantum problem and particularly Bohr's complementarity interpretation.

Plotnitsky continues the epistemological line of thinking from Kant to Bohr and arrives at his own version, which he calls reality without realism. A catchy name for a theory. What Plotnitsky means is that it may be fundamentally impossible for us to know or even draw any conclusions about the quantum realm, basically following Bohr's interpretation, which, in turn, drew originally from Kant.

Plotnitsky is very critical of ontological approaches to the quantum mystery, particularly those that try and mathematise the metaphysical structure of the world. This criticism is correct, in itself, though not because, as Plotnitsky asserts, this metaphysical structure is inherently unknowable. As we're arguing here, it's really because of the fact that such approaches presuppose the straightforward application of the truism of noncontradiction to this ontic structure.

They already pre-define too much about the world. Plotnitsky makes no comment, as far as I'm aware, on Hartmann's critical ontology.

Plotnitsky's reality without realism theory is correct from an epistemological perspective and I would agree with it, that is, based on our traditional application of the truism of noncontradiction and when the aspiration is for a priori knowledge about the ontic structure of our world.

Plotnitsky only dismisses ontological attempts to a priori solve the quantum mystery and efforts to mathematise the ontic structure of our world. He doesn't seem to realise the problem quantum discontinuity actually creates for *all* methods of analyses—logical, ontological *and* epistemological—that aspire to a priori knowledge about the ontic structure of our world. He maintains the validity of epistemological approaches, such as his own, because he continues naturally to presuppose the validity of the application of the truism of noncontradiction to the otherwise unknowable ontic structure.

The only real error Plotnitsky makes in his line of thinking is to not recognise the uncertainty that quantum discontinuity inevitably raises with respect to our application of the truism of noncontradiction. In spite even of recognising that the emergence of causality from no-causality represents the most plausible explanation for quantum interaction, Plotnitsky seems unable to realise the deeper implications of his epistemological approach, and specifically its reliance on the presupposed a priori certainty of our application of the truism of noncontradiction.

Again, its very plausibility, the fact that the limit of the phenomena is defined by the relationship between spatiotemporal discontinuity-continuity, and that this relationship could possibly precede our application of noncontradiction renders the straightforward nature of this application no longer an a priori certainty in our world. The possibility of this scenario means, by its very definition, that the application of the truism of noncontradiction cannot be taken as a priori in our world.

Because of this uncertainty, Hartmann's critical method of ontology provides arguably the best approach to understanding the relationship between quantum discontinuity and the continuous causal structure as representing the starting-point for our world, but of course, without aspiring to a priori knowledge, which this discovery renders impossible anyway.

For Plotnitsky, the evidence points toward the emergence of causality from no-causality, but he can't definitively draw that conclusion because of his epistemological approach and his continued adherence to the traditional a priori method. Plotnitsky can only speculate about this starting-point for our world but must conclude that it's impossible to know with certainty. He doesn't seem to realise that the very foundation itself for that a priori method can no longer be taken as an absolute certainty thanks to the discovery of quantum discontinuity and, perhaps ironically, to the real possibility that the starting-point could be the emergence of causality from no-causality. Instead, his methodological approach virtually forces him to be satisfied that the answer must be unknowable, at least, that is, as the evidence stands at the moment.

Without even appearing to realise, Plotnitsky has already effectively predefined the ontic structure that he is attempting to figure out. It doesn't matter that he concludes that this structure must be inherently unknowable—essentially because it must provide the origin of contradiction itself. He comes to that conclusion ultimately because the starting-point for his analysis is the truism of noncontradiction, and he aspires to a priori knowledge of that same starting-point.

Kant had originally identified this truism as the one thing we could be absolutely certain of with regard to the ontic structure of our world. It was this truism that Kant took as the starting-point for his transcendental research project. The original dilemma for Kant was how we could possibly know with certainty a real starting-point that appeared to derive from a contradiction. The solution for Kant was to conclude that the answer must be a priori unknowable, at least for us, because of the limitation of our own cognitive make-up. Plotnitsky's reality without realism theory can be interpreted as a continuation of that project.

When it comes to the history of ideas about quantum interaction, there is probably no better place to start than Kant's *Critique of Pure Reason* (1781/1787). As I pointed out before, the debate between Einstein and Bohr revolved around the different philosophies of Hegel and Kant, while Hegel's philosophy itself evolved as a reaction against Kant's first *Critique*, and even more specifically, Kant's conclusion that the ontic structure of our world must be unknowable, ultimately because of the enigmatic existence of contradiction.

Six

Kant's Unwitting Error

I like to think that Kant, if he'd been around for the discovery of quantum discontinuity, might have been astute enough to realise the implications of this discovery for our application of the truism of noncontradiction and for a priori knowledge.

For Kant, the very possibility of metaphysics was dependent on our ability to attain a priori knowledge of the world. Because metaphysics, by its definition, transcended the physical, experience alone could not be relied upon for knowledge of it. As far as Kant was concerned, the existence of metaphysics as a field of study required a priori knowledge of our world.[1] The starting-point for such knowledge, the thing that made it possible in the first place, was the law of noncontradiction as a self-evident truism.

Kant always took the law of noncontradiction as a logical truism. As the starting-point for a priori knowledge of our world this law had to be a priori itself: it could in no way be dependent on experience for its initial validity. Kant's unwitting error—like virtually the entirety of Western thinking—was to conflate the real application of noncontradiction as a fundamental ontological law in our world with the idea of it as a self-evident truism.

Once we're reasonably able to conclude that the relationship between quantum discontinuity and the continuous causal structure of our world could represent the appearance of the real emergence of causality from no-causality, then the straightforward application of the truism of noncontradiction to our world can no longer be taken simply as an a priori certainty. The initial connection of a priori knowledge to our world becomes uncertain. This is because, as the effective starting-point for our world, this relationship would have to precede absolutely everything, including the application of this truism.

The plausibility of this scenario is something that nobody seems to have yet picked-up on.

> *And it must be plausible if someone as expert as Professor Plotnitsky recognises that quantum interaction could be caused by the emergence of causality from randomness.*

It's significant that Plotnitsky compares the emergence of causality from no-causality to the emergence of order from chaos in John Milton's *Paradise Lost* (1667).[2] I'm not interested in the religion, but only in the fact that our experience of the world can be interpreted this way. It's not too dissimilar to ancient philosophers interpreting their world as starting from the relationship between continuity-discontinuity, Newton and Leibniz's contrasting views on the world as being continuous and discontinuous, and even to Hartmann understanding the most fundamental categories to be oppositional relationships, starting with continuity-discreteness.

Today we speak of a binary world. The trouble has always been how to reconcile metaphysically this concept of a binary world with the truism of noncontradiction as a fundamental ontological law and the first law of logic.

If the starting-point for everything is the emergence of causality from no-causality, then this relationship, by definition, would literally have to exist before anything else, including any knowledge, or the application of any law, even the principle of noncontradiction as a fundamental ontological law of existence.

It's not difficult to draw the conclusion that the observable relationship between quantum discontinuity and the continuous causal structure of our world could represent the appearance of the emergence of causality from no-causality. Indeed, you could expect the existence of real physical randomness to appear as discontinuous in space and time from within and as part of the same otherwise causally governed world: it's this spatiotemporal quality of quantum discontinuity that allows it never to contradict the space-time continuum of Einstein's general relativity; such randomness is not just mutually exclusive to causality, but a real jointly completing part of the ontic structure of our world.

So how does Kant deal with this problem?

Well, to start, there are a few things we need to understand about Kant and particularly his *Critique of Pure Reason*.

First, his method of approach was epistemological. Kant was following in the footsteps of Descartes in the sense that he understood that the best tool we had for gaining knowledge about the world was our own rationality. Kant's main purpose in his *Critique* was to determine how human reason actually worked and to figure out the ultimate limits of that reason. If our own reason was the best tool we had for gaining knowledge about the world, then it needed to be accurately calibrated.[3]

Second, Kant was specifically interested in knowing the limits of a priori knowledge, that is to say, what we could know about the real world with certainty, starting from first principles. Again, this presupposed the a priori nature of first principles such as the truism of noncontradiction. You don't need to prove the impossibility of an object contradicting itself. In fact, as Aristotle pointed out, it's not possible to prove the truism of noncontradiction because any such proof would already have to presuppose the law it was trying to prove.

But, again, it's not the truism itself that is the problem. A contradiction will always be a contradiction. It's our application of this truism that codifies noncontradiction as a law specifically in our world. Even Kant seems not to have sufficiently grasped the potential significance of this fact. It's this application that has become questionable because of the discovery of quantum discontinuity.

Which brings us to the third and most important point here in understanding Kant's *Critique*: the explicit starting-point for Kant *was* the truism of noncontradiction. Kant presupposed this truism to be the one thing we could know about the real world, that it could not possibly be self-contradictory.[4]

> But, of course, as you're arguing, this is not really the case anymore, is it, following the discovery of quantum discontinuity?

We can now reasonably conclude from the observable evidence that the starting-point for our world could be the emergence of causality from no-causality. Because of the discovery of quantum discontinuity, these otherwise contrary qualities can no longer be taken simply as starting out in our world as mutually exclusive to each other. They really need to be understood, from the very outset of everything, as complementary, that is to say, both mutually exclusive, yet also jointly completing in the ontic structure of our world.

So, the very starting-point for Kant's epistemological analysis of pure reason is no longer an a priori certainty. Remember, the empowering mechanism for Kant's *Critique* was the certainty of first principles. This was really the driving force behind Kant's transcendental research project, and it all started from the application of the truism of noncontradiction to our world. Without the certainty of the application of this truism as the very starting-point for his *Critique*, some of Kant's conclusions may need to be reconsidered.

You also said he believed in a causal ontology.

Kant was working about a hundred years after Newton. Like everyone else by that time, Kant took for granted that the ontic structure of our world had to be governed by the law of efficient causality. As I said, this assumption was underpinned by the truism of noncontradiction and the belief that the ontic structure of our world could not possibly be both random and causally governed at the same time, which would necessarily lead to a contradiction.

It's this way of thinking that gives physicists such a headache when it comes to trying to understand quantum interaction, particle-wave duality, for example, or the quandary of whether Schrödinger's cat is really dead or alive.

We need to re-think through these problems without pre-supposing the application of the truism of noncontradiction or a causal ontology.

These assumptions of a causal ontology and the application of the truism of noncontradiction had a defining influence on Kant's thinking.

> *And this way of thinking, as you say, mixes up the truism of noncontradiction with its application as a law in our world.*

The key is not the truism itself, but it's necessary application to our world as a fundamental ontological law. If the relationship between no-causality-causality actually exists before the application of this truism, then it would never contravene the application of this law. You could expect such inherent randomness to appear from within and as part of the same world as spatiotemporal discontinuity.

The emergence of causality from no-causality is not only possible as the starting-point for our world, but it provides the simplest and most plausible explanation, based on the phenomena and our current experience of the world.

> *OK, so the starting-point for Kant's analysis was the truism of noncontradiction, and he assumed a causal ontology.*

Kant's primary aim was to put metaphysics on a scientific footing comparable, for example, to the physics of Newton and Galileo. Kant also started his academic career as a mathematician. Kant really wanted the study of metaphysics to emulate the success of physics in how it was able to describe the physical world through the a priori application of mathematics. To do this, Kant believed metaphysics needed to ensure it also applied a purely a priori method.

As I pointed out earlier, there has always been some confusion about how our mathematical descriptions actually connect to our world. Descartes and Galileo, for example, presupposed the a priori—even divine—connection of mathematical description to our world, while Newton understood the ultimate fallibility of any such descriptions and their necessary reliance on experience. Kant was in the former camp with Descartes and Galileo, and his critical project was aimed at calibrating our a priori connection to the world, again, starting from the truism of noncontradiction.

Even by Kant's time, metaphysics was struggling for credibility. It was already the age of modern science, and the smart money was on science to be able to discover everything worth knowing about the world.

Ironically, perhaps, Kant believed that the problem with metaphysics was that it had become too bogged down in its own dogmatism. It had come to assume too much about the world. For Kant, the first step was to get rid of all the dogmatic assumptions we made about the world.

That sounds familiar.

Much of Hartmann's critical method was inspired by Kant. That's why Hartmann referred to his ontology as 'critical'.

Following Kant, it was commonly presupposed by philosophers that ontology had basically become a dead field of study. Hartmann started out by arguing against this way of thinking and arguing that Kant didn't actually do away with ontology at all, but only ontologies that took the relationship between thought and reality for granted.[5] Modern examples of this are attempts to mathematise the ontic structure of our world; the same thing that Plotnitsky argues against.

Hartmann's point was that Kant didn't go far enough in his *Critique*: he managed to work out how our cognition worked and basically how we understood the world, but Kant didn't actually get to the reason why we understood it that way. According to Hartmann, in order to do that, we needed to carry out an even more radical critique of the ontic structure of our world.

> *And in order to do that we have to set aside first principles, especially the law of noncontradiction. As you said before, according to Hartmann, Being simply 'is what it is', and we shouldn't assume anything about the ontic structure of our world.*

Hartmann recognised the potential danger of taking the initial application of noncontradiction to our world for granted. By presupposing the validity of first principles such as the truism of noncontradiction, we already pre-defined certain things about the ontic structure of our world. The truism of noncontradiction has always seemed so natural and so self-evident that even a genius on the level of Kant could take it for granted.

The trade-off is that we can no longer hope to achieve a priori knowledge of the ontic structure of our world. The best we can ever do is draw theories based on the evidence, which basically conforms to the modern scientific level of proof, anyway. This was the key point about Hartmann's critical method of ontology, and it's what really sets it above other philosophical approaches to the question of Being, including the phenomenologies of Heidegger and Husserl and even the logic of contemporary analytic philosophy.

OK, so what did Kant say in his Critique?

The pivotal point about Kant's critique of human rationality was the distinction he made between, on the one hand, our sensibility, our senses, providing us with sensory data from the external world, and on the other, an inbuilt faculty of understanding actually enabling us to make sense of that data. Kant referred to these two parts of human cognition as the separate stems of understanding and sensibility. For Kant, objective knowledge of the world derived from the synthesis of these twin cognitive stems. It was this idea of an inbuilt faculty of understanding in particular that was really novel and ground-breaking about Kant's *Critique*.

Kant basically turned our way of thinking about human cognition on its head. Instead of simply being passive recipients of the experiences we perceive, Kant effectively made cognition an active originator of that experience.

Is that what Kant meant by his Copernican revolution?

There has been quite a bit of conjecture and misunderstanding as to what Kant was actually on about with this idea of an inbuilt faculty of understanding. For one thing, Kant himself never actually referred to it as his Copernican revolution. What he did was compare his method to Copernicus, who, when confronted with the seemingly impossible problem of trying to work out mathematically the motions of the planets and stars, fundamentally changed our perspective of things and had the Earth rotate while leaving the stars at rest. Instead of having our cognition conform to experience, Kant started out by presupposing that our experience of the world must conform to our cognition.[6]

Kant never stopped believing in the reality of the external world. His version of idealism is not like Bishop Berkeley's, for example, who basically argued for the impossibility of the material world.[7] If you link Kant to this way of thinking, then you need to go back and read his work more closely.[8]

The easiest way to get your head around Kant's notion of an inbuilt faculty of understanding is to interpret it simply as a necessary part of our evolutionary make up. It's something that we all share in common, and that has evolved within us: it's really the thing that ultimately defines us as uniquely human. Kant was around in the century before Charles Darwin.

Think about it, we must have something built into us that enables us to pull all the sensory data together. Otherwise, how could we collectively make sense of the world? And what would differentiate our experience of the world from the way a dog, a chicken or a bumble bee experience the world. It can't be just our sensory apparatus alone that defines our differing experiences of the world.

> Didn't Kant say he was woken from his dogmatic slumber by David Hume?

The British philosopher David Hume had argued, quite convincingly, that there was a problem with our metaphysical understanding of causality. Hume pointed out that we couldn't have a priori knowledge of cause and effect but could only gain knowledge of causality through our experience of the world. Hume was an empiricist, which meant he believed basically that all knowledge of the world originated from experience—that is to say, apart from self-evident truisms, such as the law of noncontradiction. For Hume, our mind started out like a blank sheet of paper and everything in our mind had to have its origin first in our experience of the world.

Again, Kant's epistemological approach was concerned with what we can know with a priori certainty about the real world. For Kant, Hume's argument created a problem, and not just for metaphysics, which, remember, traditionally dealt in a priori knowledge, but also even the fundamental laws of physics that Galileo and Newton had helped codify. Put simply, without a priori knowledge of causality, Kant concluded that we couldn't have a priori knowledge of anything.

Kant identified two forms of a priori knowledge: referred to as analytic judgements and synthetic a priori judgements. Analytic judgements are true by their own definition, such as the truism of noncontradiction or that all bodies take up space. Analytic judgements are considered trivial in the sense that their validity can't really be seriously questioned, even by someone as sceptical as Hume. Synthetic a priori judgements, on the other hand, though necessarily true, still require further evidence for their validity. Causality is an example of this and so is mathematics when applied to the physical world.

The problem that Kant's first *Critique* addressed was how to guarantee the a priori status of synthetic a priori judgements without having to refer to experience. Hume argued that we couldn't.

This was one of the key functions of Kant's faculty of understanding. By creating an inbuilt source of knowledge about the world, Kant could ensure the a priori status of synthetic a priori judgements: if our faculty of understanding served to order our experience of the world, then we'd have to have prior knowledge of such understanding, before any possible experience.

To have a priori knowledge of the real world, our experience of it had to conform to our own cognition. Our understanding of causality, for example, and of space and time, had to originate from within us and not simply from our experience of the world. In this way, Kant made it possible to maintain a priori knowledge of causality and rebut Hume's scepticism.

As Kant explained, analytic judgements derived their validity directly from the truism of noncontradiction. Synthetic a priori judgements, on the other hand, need to derive their validity from somewhere else, namely our faculty of understanding. Synthetic a priori judgements are still necessarily subject to the law of non-contradiction, but they do not acquire their initial validity directly from this law.[9]

The point to keep in mind, however, is that Kant's argument itself for an inbuilt faculty of understanding was predicated on the presupposed application of the truism of noncontradiction. Put simply, the fact that such a faculty would necessarily have to come before any experience to ensure its a priori status, that is to say, in order to avoid falling into an infinite regress, presupposed the straightforward mutual exclusivity of the relationship between our understanding and experience. Kant's synthetic a priori judgements still find their ultimate origin in the application of the truism of noncontradiction to our world.

Kant argued that our inbuilt faculty of understanding was a key ingredient in our acquisition of objective knowledge of the world. It was specifically the synthesis of this inbuilt faculty with our sensibility, the sensory input from the external world, that for Kant actually served to define the concept of objective knowledge.

But, again, remember, it all started for Kant with the a priori application of the truism of noncontradiction to our world. This is really the key point. No matter how complicated things might get, the simple fact remains that Kant mixed-up the idea of this truism with its real application in our world. Noncontradiction is certainly an a priori fact requiring no evidence, but how that truism initially applies in our world as a fundamental ontological law can no longer be beyond question. The real possibility that the starting-point for our world could be the emergence of causality from no-causality, effectively the complementary relationship between spatiotemporal continuity-discontinuity, renders the very starting-point for Kant's *Critique* questionable, and no longer simply an a priori certainty.

So, what does this mean for Kant's Critique?

It's not my purpose to work out the implications of this realisation on Kant's work. That's too big a job for me to deal with here. Really, my aim is merely to convince you that the a priori application of the truism of noncontradiction to our world has become problematic with the discovery of quantum discontinuity. Suffice to point out, for example, that Kant's very idea of cognition as being defined by the separate stems of understanding and sensibility presupposed the application of noncontradiction: the mutual exclusivity of these stems was ultimately rendered necessary because of the application of the truism of noncontradiction as a fundamental ontological law. This then led to the defining of objective knowledge in terms of the synthesis of these separate stems and the separation of phenomena from noumena.

It's this separation of phenomena from noumena that was at the heart of Bohr's complementarity interpretation of quantum interaction and also Plotnitsky's reality without realism theory.

Kant also claimed that he was woken from his dogmatic slumber by what he referred to as the antinomy of pure reason, and what he cryptically described as the contradiction of reason with itself.[10] This dilemma came to be expressed primarily in Kant's four cosmological antinomy, the first antinomy being, perhaps not surprisingly, the discontinuity-continuity of space and time.

As I've pointed out, there has always been this understanding that the ultimate starting-point for our world derives from the relationship between discontinuity-continuity. In Kant's time, it was expressed most cogently in the competing theories of Newton and Leibniz. Aristotle had originally codified the truism of noncontradiction as the first law of logic and as a way of understanding this relationship as necessarily starting out as mutually exclusive. This was the starting-point for Western metaphysics, and we've been trying to reconcile this seemingly self-evident starting-point ever since. In this regard, our efforts to come to terms with the meaning of quantum interaction, and our pursuit of a mathematical theory of everything is just the continuation of this age-old problem.

For Kant, the only way to reconcile these competing perspectives of our spatiotemporal world was to separate the phenomena from the noumena. Objective knowledge of phenomena derived from the synthesis of our twin stems of understanding and sensibility; phenomena had to be ultimately part of the world we experienced with our senses in order for such phenomena to be able to derive from this synthesis. We may have been able to think about noumena, but because we didn't actually experience noumena through our senses, we couldn't possibly have objective knowledge of them.

Our knowledge of causality, for example, existed as part of our inbuilt faculty of understanding. It was the existence of this faculty that made it possible for us to have a priori knowledge of causality beyond mere experience. This was Kant's answer to Hume. It was not just through our experience that we gained knowledge of causality, but also through our necessary faculty of understanding. Causality was an unavoidable ingredient of the world, and we acquired a priori knowledge of it because it was built in to our very understanding itself.

The problem arose, however, when philosophers tried to reconcile this necessary causality and continuity with the necessity also for the world somehow to be fundamentally discontinuous (i.e., random in space and time). It was this antinomy and the apparent impossibility of ever knowing a priori the source of such apparent contradiction that was central to Kant's separation of phenomena and noumena.

The existence of such antinomy has really always been the original problem of metaphysics, the idea of the world starting out as somehow both discontinuous and continuous. Aristotle himself recognised this fact. It has been the enigmatic existence of this original antinomy, the proverbial elephant in the room, that has plagued Western metaphysics since Aristotle, aggravated by our perennial aspiration for a priori knowledge of the real world.

The significant difference today is that, with the discovery of quantum interaction, evidence of such a relationship between spatiotemporal discontinuity-continuity has been found to define the physical limit of our world. The idea of it is no longer simply the product of our metaphysical reasoning. With the discovery of quantum interaction, it's this relationship that has been found to represent the physical limit of what we will most likely ever be able to measure in our world.

It is only ultimately our presupposed need to satisfy the truism of noncontradiction beyond this physical limit that drives us to assume that there must be more to the real world. If the starting-point for everything is the emergence of causality from no causality, then this relationship of spatiotemporal discontinuity-continuity would define not just the limit of the phenomena, but also the very limit and effective starting-point for the real world itself. Because the application of the truism of noncontradiction to our world can no longer be an a priori certainty, and, really, doesn't need to be, then this relationship clearly provides the simplest and the best explanation for the starting-point and ontic structure of our world.

Kant assumed the need for a causal ontology and, basically, set aside the solution to the apparent contradiction itself between Newton and Leibniz's perspectives by asserting that such a solution could only exist in the noumenal realm. It was thus beyond any hope of ever being objectively known. Meanwhile, Bohr and Plotnitsky have taken a similar stance with regard to the contrary relationship between quantum discontinuity and the continuous causal structure of the world. Although this relationship itself is part of the phenomena we experience, and we can have objective knowledge of it, according to Bohr and Plotnitsky, the ultimate origin of its contrariness is necessarily beyond our experience. In other words, it must be part of the noumenal realm and beyond any possibility of objective knowledge, even through the application of mathematics.

This whole line of thinking, though, presupposes the simple mutual exclusivity of phenomena and noumena and ultimately the a priori necessity of the application of the truism of noncontradiction to the real world. Once the certainty of this presupposition is brought into question, this whole way of thinking about phenomena and noumena becomes questionable, even simply unnecessary.

The most plausible explanation is that it all originates from the emergence of causality from no-causality. The confusion and complication come about entirely as a result of our unnecessary assumptions about the world, starting with the mix-up between our application of the law of noncontradiction and the idea of it as an a priori truism.

OK, so what about Hegel?

Seven

Hegel Made the Same Unwitting Error

The risk in talking about Hegel's philosophy is getting too bogged down in the detail, even more so than with Kant. Hegel may be the most difficult of all the great philosophers to get your head around, at least Bertrand Russell seemed to think so. It was primarily Russell's summation of Hegel's philosophy, along with Frege's, that helped define Anglo-American attitudes to Hegel for most of the twentieth century. Hegel's ideas certainly didn't fare well under Russell's analytic scrutiny. Understandable, I suppose, given Russell was making his assessment in the wake of Hitler, two world wars and the rise of Marxism, not to mention, Russell's well-known pacifism and Hegel's apparent efforts to justify war.

It's fair to say that Hegel was one of the more divisive figures in modern philosophy. Even in his own time, his ideas tended to be either vilified or lauded. Schopenhauer, for example, described his writings as pure nonsense, an extravagant maze of words and a monument to German stupidity.[1] To be fair, Schopenhauer was a professional rival of Hegel's. I suspect, though, that too many philosophy students, who've had to wade through Hegel's work, might tend to agree with this summation.

Still, Hegel did manage to capture the intellectual imagination of his time and even became popularised beyond academia, with workers in the streets of nineteenth century Europe being known to sprout Hegelian ideas.[2] Then, of course, there's Marx who famously put his own materialist and communist spin on Hegel's theory of historical development. I'm assuming you know how that worked out.

So, what's your opinion of Hegel?

I find his writings tedious, to be honest. But then, I find most philosophy tedious. Russell's analytic logic also did a thorough job of sucking the life out of things. It's not unlike how Kierkegaard described Hegel's work, as essentially overlooking what it really means to live.[3] Meanwhile, Schopenhauer and Kierkegaard are rather too pessimistic for my taste. I'm more fascinated, you could say, with the development of philosophical ideas, how they came about, rather than the ideas themselves.

These days, I tend to read philosophy through the lens of this mix-up with the application of noncontradiction. I can't help thinking that most of the great philosophers were unwittingly misguided from the start. The real genius for me lies in how Hegel and co still managed to formulate coherent metaphysical theories in spite of this misunderstanding.

Hegel really needs to be understood in the context of Kant. It's not surprising, I think, that there has been a push in recent times against the old analytic criticism characterised by Russell and to reassess Hegel's work, to interpret it in relation to Kant's critical philosophy and to see Hegel more as a continuation of that philosophy. Remember, Kant's aim was to put metaphysics on a scientific footing comparable to the physics of Galileo and Newton. Hegel's writings can be understood as following on from that project.

> *Apart from his apparent justification for war, why exactly did Russell criticise Hegel's philosophy?*

The main issue, I think, for Russell was the mysticism and the religious connotations permeating through Hegel's writing. This really culminated in what Hegel described at the end of his most famous work, *Science of Logic* (1812), as the so-called 'Absolute Idea'. Not without good reason, Russell derisively attributed this to Hegel's assertion in the existence of a divine, all-knowing Being. Of course, as you might expect, Russell, who was an affirmed atheist, had convinced himself of the logical impossibility of God as a teenager, apparently after reading John Stuart Mill.

As I said, there has been a reassessment of Hegel's work in recent times as more a continuation of Kant's project of trying to put metaphysics on a scientific footing. For Hegel, logic and metaphysics essentially amounted to the same thing. Of course, the idea that metaphysics and logic are basically the same was also the position of Russell and is presupposed, as well, by contemporary analytic philosophy. It's just that, how Russell arrived at that position and what it actually meant for him, compared to Hegel, was somewhat different.

Russell's metaphysics was aligned with that of modern science, of a spatiotemporal world of moving parts, governed by efficient causality and the truism of noncontradiction, and entirely describable through mathematics. The old Aristotelian and medieval idea of a final cause for everything and a divine being had been thoroughly extirpated from this modern scientific understanding of metaphysics. Following Frege, Russell attempted to create a mathematised language that could logically, efficiently and a priori describe the world at its metaphysical level without the need to resort to divine intervention. Of course, Russell was instrumental in the development of analytic philosophy.

According to the reassessment of Hegel, on the other hand, his Absolute Idea can be understood as defining not so much a divine being as such, but pure metaphysics itself, purged of all dogmatism and false assumptions. For Hegel, a priori knowledge of our world is assured only by being able to know the whole world in its entirety. This then is the end goal of metaphysics and history.

> *How are you supposed to know everything?*

The point, I think, is that metaphysics and history strive toward this Absolute Idea through the application of logical thinking. Remember, logic and metaphysics for Hegel are the same thing. The concept of an Absolute Idea ultimately allows Hegel to overcome the limitation Kant had put on our knowledge of the real world. Remember Kant's division between phenomena and noumena?

> *Phenomena were things that we could experience with our senses; noumena were things beyond experience.*

Kant arrived at this separation as a way of overcoming the metaphysical problem of contradiction, how to accommodate a world that appeared to be both intrinsically continuous and discontinuous. We can experience this continuity and discontinuity as mutually exclusive phenomena, but we can't actually experience the source of the contradiction itself. Somewhere along the line there must be a solution, but that solution had to be part of the noumenal realm and beyond the possibility of objective knowledge.

Can I ask how modern science deals with this problem?

Well, modern science doesn't really dwell too much on the problem of contradiction. And it doesn't bother too much with a starting-point, either. We can include Russell and analytic philosophy in this. Sure, they presuppose the straightforward application of the truism of noncontradiction to our world—they assume it logical heresy to do otherwise—but they put their faith in mathematics to ultimately describe the physical world. As long as our spatiotemporal world of moving parts continues to function according to the principle of efficient causality, modern science takes it for granted that it should be possible to describe it entirely through mathematics. The task for modern science is to discover the mathematical laws that govern the real world and that coincide with our experience of it.

I think both Kant and Hegel were fundamentally different to this because they explicitly took as their starting-point the truism of noncontradiction, as the one truth about the real world that we couldn't possibly doubt (apparently), that the world could not be self-contradictory. Basically, they followed the traditional philosophical method of starting from first principles and built up their picture of the world through a priori analysis. It doesn't matter that Kant considered the ultimate source of contradiction to be unknowable because he still presupposed its validity. This presupposition effectively served as the ontic starting-point for

Kant. Even if it lay ultimately in the noumenal realm and was objectively unknowable, such a starting-point was still dictated by the truism of noncontradiction and mutual exclusion.

Of course, modern science also presupposes contrary relationships to be mutually exclusive in the real world, but it doesn't presume this truism to be its starting-point, at least not explicitly. If anything, it seems to take efficient causality to be the physical starting-point, which probably makes sense when the real starting-point can be understood as the emergence of causality from no-causality. As I said, though, modern science doesn't really worry too much about a clear starting-point so long as our experience continues to affirm that our world is causally governed. Again, this allows modern science to continue hoping that it'll all be explainable through mathematics.

> *You mean that we can hope one day to arrive at a theory of everything?*

The problem with the discovery of quantum discontinuity is that it has brought this neat way of understanding the world into question: the spatiotemporal randomness of quantum interaction seems to contradict the law of efficient causality and thus to threaten the entire theoretical edifice that science has built up over the past several hundred years.

The fact that modern science doesn't clearly define its starting-point creates a metaphysical problem, however. Kant, Hegel and traditional metaphysics understood the basic need for a starting-point. Their problem was that they mixed up the application of noncontradiction as a fundamental ontological law with the idea of it as a logical truism.

The reason we need to be clear about our metaphysical starting-point is because it may actually have a defining influence on the way we do our analyses. Even though modern science doesn't dwell on the matter, it does still effectively take as its logical starting-point the law of noncontradiction, albeit, by default. It's specifically because of the application of the truism of noncontradiction that we presume there to be no room for inherent randomness in the physical structure of the world, that is to say, the idea of 'no-causality' actually built into its real, ontic structure. According to the truism of noncontradiction, the physical world, to be more specific, the starting-point for the physical world, can't be both causally governed and random at the same time. That's why physicists try so hard to maintain the mathematical status quo and come up with a theory of everything that can maintain the universality of efficient causality in our world.

Aristotle understood that our knowledge about the world starts with our application of the truism of noncontradiction. The error Aristotle made was to mix-up the application of noncontradiction as a fundamental ontological law in our world with the idea of it as a logical truism. It's this real application that must ultimately serve as the logical starting-point for all knowledge in and about our world.

This makes even more sense once you accept the possibility that the ontic starting-point for our world could be the emergence of causality from an original state of no-causality. From our necessary location within and as part of the same world, this real starting-point can be expected to appear to us as the complementary relationship between spatiotemporal discontinuity-continuity. If this really is the case, the initial application of the law of noncontradiction would not be defined by mutual exclusion, as we've always been led to assume because of its logical truism, but both mutual exclusion and joint completion.

The problem with our modern scientific approach is that we've effectively extricated ourselves from the real world we're trying to know.

> *We need to look at everything from the perspective of human cognition?*

It's not because of some quaint idealism that wants to render everything as originating from our own cognition, but because of the simple, unavoidable fact that any starting-point for our world must also be the starting-point for any knowledge about our world. This really needs to be factored into our calculations, which modern science doesn't do very well, and come up with objective knowledge about the world which has us, more specifically knowledge itself, in and as part of that same world. This is the real metaphysical dilemma, and why the application of mathematics or logic alone is insufficient for the task.

This is basically what Kant and Hegel were trying to do, but even they failed to appreciate that the metaphysical (i.e., real) starting-point must also precede even the application of noncontradiction as a fundamental ontological law and first law of logic. We can't work out the ontic structure of the world until we've actually figured out the starting-point for it, including all possible knowledge in and about that world.

We've always taken this starting-point for granted because we've always taken it a priori to be, either directly or by default, the law of noncontradiction. The real significance of the discovery of quantum discontinuity is that it has brought this metaphysical starting-point into question.

By taking the ontic structure to be completely describable through mathematics, modern science fails to realise this fact. We've got to work out the ultimate starting-point for everything before we can hope to do anything else, including describe the world mathematically, and it needs to be done ontologically because it is specifically an ontological problem, that is, a question for 'first philosophy'.

> OK, I've got that, but you still haven't explained how Hegel went from the truism of noncontradiction to his Absolute Idea.

Well, as I just said, Hegel, like Kant, took the starting-point for his theorising to be the truism of noncontradiction—effectively, its straightforward application to our world. The key dilemma was how to overcome the existence of contradiction in the metaphysical world. The epiphany for Hegel came with the introduction of time, that is, different moments, into the ontic mix.

Hegel sat back in his armchair and identified the ultimate metaphysical tension to be between Being as such and nothing. But, as he surmised, these two concepts had to represent different 'moments' in the ontic framework. According to the law of noncontradiction, Being and nothing can't possibly exist in the same object at the same time. Metaphysical completion had to initially come about through the synthesis of Being and nothing and then emerge (so to speak) as becoming.

The logical starting-point for everything had to be the synthesis of Being + nothing = becoming.

> This was the dialectic?

For Hegel, the way to overcome contradiction was with the introduction of the third concept of 'becoming', which was made possible by the introduction of distinct moments. This then became the blueprint for the dialectic, thesis + antithesis = synthesis, and the starting-point for our logical thinking about the world.

It's how Hegel overcame the Kantian problem of unknowable noumena. Metaphysics equated to logic, and logical thinking about the real world was structured in the form of this dialectic. Historical development came about through the continuous cycling of thesis + antithesis = synthesis.

The concept of Being only had logical meaning in the context of its relation to nothing. Such contrary relationships for Hegel were not simply mutually exclusive, but actually started out from the beginning as jointly completing aspects of the real world.

Like what you're saying, then?

Not really. Hegel, like basically everyone else, never doubted the straightforward application of the truism of noncontradiction to our world. This was his logical starting-point. Like Kant, Hegel began by taking for granted the one thing we could supposedly never question, that the real world could not possibly be self-contradictory. He never gave up that assumption.

Hegel tried to figure out the consequence of a starting-point of Being-nothing while always presupposing the necessary application of the truism of noncontradiction. He deduced their dialectical joint completion in the concept of becoming.

You could argue that Being-nothing can also be conceptualised as preceding the application of the truism of noncontradiction in our world. In effect, this relationship is the metaphysical equivalent of the physical relationship between causality and no-causality. The former, at the very least, must contain the latter. To be coherent, the concept of 'Being' requires causality, just as 'nothing' can be defined, at its barest minimum, as the state of no-causality. Put another way, causality defines the difference between Being and nothing. You could argue, I suppose, that randomness (i.e., no-causality) is simply another state of Being—rather than nothing itself—but this makes little practical sense when the definition of randomness *is* no-causality. It's a little obtuse to deny such a comparison when the potential payoff is the conceptualisation of Being-nothing as complementary. Understanding them as effectively the same opens the possibility of conceptualising this metaphysical relationship as inherently mutually exclusive and jointly completing in our world—a proverbial 'holy grail' for metaphysics.

Hegel's conceptualisation of Being and nothing, on the other hand, as a dialectical relationship to explain the ontic starting-point for our world was little more than a logical convenience, born from the assumed need to satisfy the truism of noncontradiction. Without this presupposition, the physical emergence of causality from no-causality makes more sense as the real starting-point for our world. This must render, also, the first law of logic as also defined from the outset by both mutual exclusion and joint completion.

Hegel just replaced one with the other, that is, mutual exclusion with joint completion.

The end product was Hegel's Absolute Idea: everything has meaning only in relation to everything else. A priori knowledge of the real world was achieved through the logical and systematic process of the dialectic, figuring out how everything relates to everything else. For Hegel, meaning couldn't exist in isolation. This holistic way of thinking has become quite popular these days. It's encapsulated, for example, in the idea of relationalism.[4]

But, a little ironically, perhaps, such holism doesn't actually give up the supposed need for ultimate mutual exclusion in accordance with the necessary application of the truism of noncontradiction. It logically satisfies this need in the presumption itself that a choice has to be made between mutual exclusion and joint completion. This is at the heart of Hegel's dialectic. The synthesis of thesis and antithesis, their joint completion, is taken as inevitable; Being and nothing must lead logically to becoming; meaning can only derive from the relationship of a contrary with its opposite.

In this way, Hegel supposedly made it possible to know a priori the ontic structure of our world and even the ultimate source of contradiction itself, that is to say, by starting with the logical relationship between Being and nothing. We didn't have to settle for Kant's idea of a noumenal realm and the ontic structure of our world being a priori unknowable.

In the same way, the idea of a real, ontic world only had meaning for Hegel in its contrary relationship to appearance. This relationship could and needed to be worked out dialectically through the application of logic. For Hegel, this gave logic an ontological dimension: it related directly to real content. Thoughts were abstract entities just like numbers were presumed to be in theoretical physics and their application of mathematics. The problem for Hegel, as with Frege and analytic philosophy, was how to go from abstract entities to concrete entities, from concept to object—the answer was through logic. This made logic, like mathematics, capable of accessing the structure of the real world.

This way of thinking represents the overwhelming position of both contemporary philosophy and science, that the real world should be entirely describable through logic and mathematics. It's only very much a minority that suspect there may be a problem with this method of approaching the metaphysical world.

Although there is a basic difference between how Hegel arrived at this way of thinking and how analytic philosophy came to it, and regardless of whether you ascribe to one or the other of these, or both, the solution to the metaphysical problem of quantum discontinuity is taken inevitably to lie with mathematics and logic.

> *And, you're arguing that the overwhelming majority have it wrong!*

Only to the extent that they haven't sufficiently grasped the implication of quantum discontinuity on our understanding of the metaphysical starting-point for our world, and for such a priori methods of analyses. As I said at the beginning, the solution to the quantum mystery, the real structure of our world and how we apply the idea of noncontradiction as the first law of logic all amount to the same problem.

I agree that Being and thinking can be taken as mostly the same, but you can't arrive at the ultimate ontic starting-point for everything through a priori analysis alone. And, you can't simply take this starting-point for granted based on the certainty of the truism of noncontradiction. It's not it's truism or apriorism that codifies noncontradiction in our world, but its real application as a fundamental ontological law. At some point there has to be a reference back to the phenomena. This is the significance of the discovery of quantum discontinuity and specifically its relationship to the continuous causal structure of our world as defining the limit of our physical world.

Hartmann basically began his critical ontology by arguing the same point, though not specifically with regard to quantum discontinuity, but with the similar age-old relationship between the concepts of form and matter. This is not a new problem, and it didn't start with Hegel or Kant. It has its roots in ancient philosophy and what is commonly referred to as Parmenides' 'identity thesis': 'thinking and Being are one and the same'.[5] The Hegelian system and even most analytic philosophy take this thesis of the equivalence of the rational and the real for granted.

So, what did Hartmann say about it?

Hartmann pointed out that the essence of form has been presupposed, since ancient times, to be its logical structure. Because we can't gain direct a priori knowledge of Being, and as the form of Being and the form of thinking seem to be essentially the same, there has always been a temptation to assume that logic should be able to provide a path to knowledge of Being itself. As logic provides the rules of thinking, it makes sense that pure logic should be able to immediately grasp the form of Being, that is to say its ontic structure, without even bothering with experience.

The history of Western philosophy is signposted by this temptation to reduce metaphysics to logic and to take the structure of thinking and the structure of Being as simply one and the same. This is where contemporary philosophy overwhelmingly finds itself today, and it is also the metaphysical position of modern science, albeit with mathematics replacing the traditional role of logic.

As Hartmann pointed out, however, approaching the problem this way makes the solution exceedingly easy. The mathematics and the logic may be very complicated, and even appear abstruse to the average layperson, but their ultimate validity is rarely doubted.

The basic error in this approach is that it effectively presupposes the formal structure to come before matter. As Hartmann also pointed out, if it wasn't for the problem of matter, indigestible like a bad conscience in the background, it would have meant the complete hegemony of logic (and mathematics) in our metaphysical understanding of the world.[6]

I don't follow.

At the end of the day, putting form before matter boils down to a value judgement. It naturally assumes that the logical structure must come before the physical world.

A case in point is exactly what is being argued here. We presume that the principle of noncontradiction as the first law of logic, and even as a fundamental ontological law, must precede the physical world that we apply it to. Of course, simple armchair logic dictates that the truism of noncontradiction, in itself, is an absolute certainty that can in no way be circumvented. That's been the starting-point for logical thinking since Aristotle. But again, it's not what we're talking about. We're really talking about the application of noncontradiction as a fundamental law in our world. This has been the mix-up in Western thinking since Aristotle. If the starting-point for everything is the emergence of causality from no-causality, then this physical relationship would have to precede the initial application even of the law of noncontradiction in our world. In this case, matter would have to precede form!

It may be argued that Hegel dug deeper than Kant and everyone else before him, but he did so a priori and by taking as his logical starting-point the truism of noncontradiction. Like Kant and almost everyone else, Hegel seems to have taken it for granted that this truism, that is to say, its application as a fundamental ontological law, must come before the physical world it's applied to. Because he already had the law of noncontradiction in place as an a priori first principle, Hegel homed in on the dialectic as a way of accommodating the existence of contradiction in our real world. He then identified the relationship between Being and nothing as the ontic starting-point for everything, and concluded that becoming must be the logical synthesis of this fundamental relationship, and thus the ultimate driving force behind everything in our world.

The discovery of quantum discontinuity requires a reference back to the phenomena without simply presupposing the a priori application of the law of noncontradiction. This discontinuity is a unique phenomenon because it marks the acknowledged limit of the observable world. It's for this reason that it can also be taken as representing the appearance of the real physical starting-point for our world. From a metaphysical perspective, and drawing at least partially from Hegel, this starting-point may even be conceptualised in terms of the complementary relationship between Being and nothing, while physically, as we've been arguing, it's characterised by the real emergence of causality from no-causality.

The relationship between quantum discontinuity and the continuous causal structure of the world can be taken as representing the appearance of this physical starting-point from within and as part of the same world.

The most significant outcome of this realisation is that this initial relationship between no-causality-causality, that is, Being and nothing, can be understood as not simply mutually exclusive, as is overwhelmingly presupposed by modern science, or jointly completing, as was presumed by Hegel, but is inherently complementary from the outset. In other words, as the real starting-point for all knowledge in and about our world, this relationship can be taken as fundamentally both mutually exclusive and jointly completing.

Eight

Again, the Problem Starts in Western Metaphysics

We could spend forever going through the history of ideas and considering those ideas through the lens of this reinterpretation of our application of noncontradiction. Kant and Hegel serve our purpose here because they both explicitly set out to redefine Western metaphysics, taking as their aspiration a priori knowledge of our world and their logical starting-point the application of the truism of noncontradiction.

Also, their differing philosophies have been so influential in helping to shape contemporary Western thinking. As I pointed out at the beginning, the respective positions of Bohr and Einstein regarding the implications of quantum interaction on our knowledge of the world can be understood as reflecting the opposing ideas of Kant and Hegel. From an epistemological perspective, the question of whether we can have access to the ontic structure of our world presents itself as the key issue.

Most analytic philosophers may disagree with Kant and Hegel, but they still invariably see this epistemological question as the underlying issue and the application of the truism of noncontradiction to our world as an a priori certainty. They see logic as being able to provide the answer to this question, just as mathematics is commonly taken to hold the key to understanding the structure of the physical world. Like Kant and Hegel, analytic philosophers invariably strive for a priori knowledge about our world, satisfied in the inevitable logic of its structure.

This whole line of thinking, however, whether you adhere to the analytic method or follow the heritage of Kant and Hegel, presupposes the necessary application of the truism of noncontradiction. To do otherwise is seen as logical heresy. Its application appears so self-evident. Again, it's not the truism itself that is problematic, a contradiction is a contradiction; it's our assumption as to how this truism initially applies in our world. We mix-up this application with the idea of noncontradiction as a logical truism. By taking this application for granted, we presuppose too much about the world beyond what can be observed and measured, regardless even of whether you believe it possible to access its ontic structure.

The central idea of this book is really quite simple. When it comes to philosophy, it's easy to get bogged down in the detail and miss the point. The point here is that our age-old application of the law of noncontradiction as a straightforward truism can no longer hold-up in the wake of the discovery of quantum discontinuity.

The mere possibility that the relationship between spatiotemporal continuity-discontinuity that defines the limit of observable phenomena could represent the appearance of the real emergence of causality from no-causality, and the effective starting-point for our world, means that the straightforward application of the truism of noncontradiction can no longer be taken simply as an a priori certainty. A priori knowledge, by definition, is presupposed to be self-evident; it doesn't require any reference back to the phenomena. There can be no doubt about it.

The idea of noncontradiction as a logical truism is beyond doubt, but that is not what actually defines noncontradiction as a fundamental law in our world: what defines it is its real application specifically as a law in our world. This is the mix-up Western metaphysics has always made. The discovery of quantum discontinuity and the possibility that the real starting-point for our world could be the emergence of causality from no-causality renders the initial application of this law open to question. This is because, as the starting-point for absolutely everything in our world, such a relationship would have to precede the initial application of the truism of noncontradiction.

The emergence of causality from no-causality is plausible and has been recognised by leading theorists and commentators on quantum physics. What has not yet been realised is the implications of this scenario, not just on the discovery of quantum discontinuity, but on our metaphysical understanding of the world.

The implications of this realisation are what we need to come to terms with in regard to our discovery of quantum interaction. The only implication being argued here is that the initial application of the law of noncontradiction needs to be understood not merely in terms of the mutual exclusion of spatiotemporal continuity-discontinuity, but both the mutual exclusion and joint completion of this relationship in our world—that is to say, its inherent complementarity. This is because, if this relationship can be taken as representing the appearance of the real emergence of causality from no-causality, and as the effective starting-point for our world, then it would have to exist before any application of the law of noncontradiction in our world.

What does this mean for a theory of everything?

I'll leave that for the physicists and mathematicians to figure out. As I said, it shouldn't really change quantum mechanics much, at least in the way quantum mechanics is able to describe and predict the effects of quantum interaction in our world.

It should have an influence, however, on how we conceive the ontic structure of our world and, for example, on how we understand quantum objects. At the heart of quantum interaction is the emergence of causality from no-causality, and specifically as it appears to us at the limit of our world as the complementary relationship between spatiotemporal continuity-discontinuity. By definition, any causality must originate from the continuous causal structure of our world. What we conceive as quantum objects, their behaviour, can't derive from anywhere other than this causal structure, and of course, its complementary relationship with quantum discontinuity. The idea of real objects, particles or waves, or some sort of hybrid mix somehow existing in space and time beyond this observable limit is a misconception based on the presupposition of a causal ontology, and ultimately, the straightforward application of the truism of noncontradiction to our world. This misconception is what we first need to come to terms with.

You're saying it's our understanding of quantum objects that needs to change? But what about the quan-

tum particles that scientists observe in their experiments?

As I said, we never actually observe quantum particles as such; we observe random quantum events, flashes of light. Scientists extrapolate the existence of quantum particles based on the accumulated behaviour of these events in their experiments. It seems perfectly natural, based on our modern metaphysical understanding of the world, to relate the cause of quantum interaction ultimately to real particles existing in space and time. This is underpinned by our assumption of a causal ontology. What we're effectively doing by this assumption is pre-defining the ontic structure of our world, it's starting-point, based ultimately on our application of the truism of noncontradiction to that structure.

The quantum interaction that leads to our conception of quantum particles is certainly real and measurable. What we need to do is rethink the cause of that interaction.

We need to rethink the metaphysical foundation itself upon which our conception of quantum interaction derives. This is perhaps a more difficult prospect to come to terms with. We're really talking about a metaphysical understanding of our world that has been built up over several centuries of successful scientific discovery. As a result of this success, it's natural to assume that science and mathematics should be able to solve the mystery behind quantum interaction.

It's also founded on two and a half millennia of taking for granted the application of the truism of noncontradiction, as essentially the default starting-point for everything.

Because we take this application as self-evident, and buoyed by the success of mathematics in being able to describe the physical world, we've preoccupied ourselves since Descartes first with epistemology and now with logic, and what we can know with certainty about our world.

Western philosophy fails even to recognise that our application of noncontradiction has become a metaphysical issue. Because of the obvious certainty of noncontradiction as a logical truism, we take the initial application of this truism as also self-evident. We're so certain in the necessary application of this truism that any questioning of it as a fundamental law tends to get dismissed out of hand. Modern philosophy mixes up the idea of noncontradiction as a self-evident truism with the application of it as a fundamental ontological law. Even when philosophers do attempt seriously to question this application, such as with quantum logic, they invariably presuppose it still to be a problem for logic, without even realising that such an a priori approach inevitably pre-defines the very thing being studied.

The end result of this age-old mix-up is the presumption of a causal ontology, a presumption that we continue to take almost entirely for granted, even in the wake of the discovery of quantum discontinuity.

Such an ingrained conception of our world is going to be very difficult to change.

> But, the theory of relativity managed to change our conception of space and time.

That was a change in thinking supported not merely by our observations of the physical world, but also by the apparent certainty of mathematics. What we're really talking about here is the metaphysical foundation itself upon which general relativity and other such mathematical theories necessarily connect to our world, that is to say, our application of the truism of noncontradiction.

To make it even more difficult, this change in our thinking can't be founded on what we've always taken to be a priori knowledge. Remember, such knowledge is not meant to require any reference back to our experience of the world for its validity. It's supposed to derive from prior knowledge, starting with first principles and ultimately the self-evident certainty of the truism of noncontradiction. This has provided the standard for metaphysics since Aristotle, and it's still the common aspiration of contemporary philosophy.

> *But you said the mathematics of theoretical physics is not a priori. It has to refer back to our experience of the world.*

The a priori function of mathematics when applied to the physical world is a common misconception. Even such seminal figures as Descartes, Galileo and Kant presupposed that our mathematical descriptions of the physical world needed to be a priori in nature. The threat to this presupposition is really what woke Kant from his dogmatic slumber when Hume claimed we could not have a priori knowledge of causality. Kant took the self-evident certainty of the application of noncontradiction for granted. The possibility of a priori knowledge, for Kant, effectively started with causality and our a priori knowledge of it.

Newton was correct to believe that the application of mathematics to our world was not a priori and didn't need to be. Mathematical descriptions must always refer back to the phenomena they are describing for their validity and always ultimately remain falsifiable. As this book is arguing, even something as seemingly self-evident as our application of the law of noncontradiction must be open to amendment based on our experience.

It's this realisation that represents the true significance of the discovery of quantum discontinuity.

This is really a problem for what Aristotle originally referred to as 'first philosophy'. It's specifically an ontological problem and one that needs to be approached first ontologically, and by setting aside our presupposition about the application of the truism of noncontradiction.

Nicolai Hartmann understood the need to set aside our application of noncontradiction in the study of the ontic structure of our world, even if he never realised the significance of quantum discontinuity on our understanding of noncontradiction as a fundamental law.

First philosophy is concerned with our foundation for knowledge, and ultimately, the starting-point for everything, including the application of any law. This starting-point can't be indubitable, as Descartes asserted, because such a starting-point presupposes our application of the truism of noncontradiction. This is why it needs to be worked out ontologically, not logically or mathematically, or through an epistemology that aspires to a priori knowledge. It's the futility of this quest for indubitability that has finally been made apparent by our discovery of quantum discontinuity at the limit of the observable phenomena and its implications for our application of the law of noncontradiction.

As a result of this discovery, it's become clear that our very foundation for knowledge itself has to be referred back to the phenomena. It's no longer possible for the starting-point to be a priori or based on indubitability! The problem of quantum interaction, the application of noncontradiction and the ultimate starting-point for our world all amount to the same problem. Until we're able to come to terms with the metaphysical implications of this discovery, quantum interaction will continue to appear an enigma to us.

And, unfortunately, Western metaphysics will most likely continue to be perceived as trivial beyond the walls of academia.

It reminds me of a recent American comedy series called *Young Sheldon*. It's about a young boy who wants to become a theoretical physicist. He's a genius, of course. His heroes are Einstein and Stephen Hawking, and his ambition is to solve the unified field theory, basically the theory of everything. It's an apt career goal for young Sheldon; I have no problem with that.

The reason I mention this show is not because of the physics, but because of the way it portrays Western metaphysics. It only touches on it very briefly, but it's quite telling in the way the show highlights the tragic conundrum that modern Western philosophy finds itself in.

Young Sheldon starts university and gets an introduction to Western philosophy: Epistemology 101.

'How do you know you're not a butterfly?'[1]

This question temporarily throws young Sheldon's thinking out of whack. If he can't even be certain he's not a butterfly, how can he know his mathematics is real. What's the point of anything?!

Sheldon does some reading: predictably, Descartes' *Meditations on First Philosophy* and also Rudolf Carnap's *The Logical Structure of the World* (1967).

The question about the butterfly is never really answered. Nonetheless, Sheldon's world is eventually brought back to equilibrium by his professor simply affirming the worthiness of the question itself and Sheldon coming to realise the worthiness of even thinking about such a question.

Unfortunately, I take this whole epistemological line of thinking as trivial! Just as I suspect the question came across to the majority of non-philosophers watching the show. Give a little smirk, think what a curious question, and get on with the serious business of living life. Just as young Sheldon did after one episode.

Actually, Sheldon's initial reaction was probably the best possible response to this question. Obviously, a butterfly's brain doesn't possess enough neurons to form a sufficiently complex dream.

But how do you know that?

First, this whole line of thinking presupposes that the proper foundation for knowledge should be indubitability, or at least, that it needs to aspire to indubitability. As I said, this presupposition is ultimately founded on the age-old certainty of the application of noncontradiction as a straightforward truism in our world. This has always provided us with the metaphysical starting-point for a priori knowledge.

Sheldon's immediate response echoes the sort of answer the Scottish philosopher Thomas Reid would have given to such a question. Reid was a contemporary of Hume and Kant and is known for the philosophical principle of common sense. Like Kant, it was Hume's scepticism that inspired Reid to write his most famous treatise, An *Inquiry into the Human Mind: on the Principles of Common Sense* (1764). The point is, there are certain things we have no choice but to take on face value, even if we can't adequately explain them; to do otherwise is simply absurd.

But what are those things?

Causality, for starters, the thing Hume so elegantly argued against. It would not even be possible to form a rational thought without causality to underpin it. Indeed, our a priori knowledge of causality doesn't even have to derive from an inbuilt faculty of understanding, as Kant argued: literally nothing is possible without the existence of causality to underpin it. Without causality there is nothing, at least beyond a state of complete randomness. Then, there is space and time, again both necessary for thought itself.

What if everything is an illusion?

Reid compared Hume's scepticism to a toy hobby-horse: something you can ride to your heart's content in private, but do so in public and they'll probably put you in a madhouse.[2]

The point is that the original foundation itself for a priori knowledge about our world, that is to say, our application of the law of noncontradiction, can no longer hold up simply as an a priori truism. The whole aspiration for indubitability that has driven modern Western philosophy since Descartes has finally been shown to be futile. Following the discovery of quantum discontinuity, we really have little choice but to re-think our metaphysical understanding of the world—or stay on the hobby-horse.

Hartmann provided us with the ontological tools to do this. The error Hartmann made was to assume that, without the certainty of noncontradiction, we needed to discover everything there was to know about the ontic structure before we could make a final judgment on it.

If we're really talking about the effective starting-point for our world, then the most obvious place to start is at the limit of that world, in other words, with quantum discontinuity and specifically its relationship with the continuous causal structure of our world. This is a relationship commonly acknowledged as defining the limit of what we will ever be able to observe and measure in the physical world.

The simplest explanation for this relationship is that it represents the emergence of causality from no-causality, at least, as that relationship must appear to us from within and as part of the same world.

The key is to realise that such a relationship, as the effective starting-point for literally everything in our world, including all possible knowledge, and the application of any possible law, must precede the application of the law of noncontradiction, even as a fundamental ontological law and the first law of logic in our world. The truism of noncontradiction remains a logical truism, a contradiction is a contradiction, but how that truism initially applies specifically in our world must also initially define it as a fundamental ontological law.

As the starting-point for our world is itself defined by a contrary relationship, then the initial application of the law of noncontradiction must be defined not merely by mutual exclusion, but by the necessary mutual exclusion and joint completion of the relationship between quantum discontinuity and the continuous causal structure of our world.

The smartest minds of the last hundred years have tried to solve the mystery behind this relationship.

It could all really be a lot simpler than we think!

ENDNOTES

Introduction

1. Arkady Plotnitsky, The Principles of Quantum Theory, from Planck's Quanta to the Higgs Boson the Nature of Quantum Reality and the Spirit of Copenhagen, Switzerland, Springer International Publishing, 2016, p.x.

2. Paula Gottlieb, "Aristotle on Non-Contradiction (Stanford Encyclopedia of Philosophy)," Stanford.edu, 2019, https://plato.stanford.edu/entries/aristotle-noncontradiction/ .

3. Tim Maudin, "The Labyrinth of Quantum Logic," *History and Philosophy of Physics*, 2018.

The Problem Begins in Western Metaphysics

1. Arkady Plotnitsky, Reality without Realism: Matter, Thought, and Technology in Quantum Physics. Cham, Switzerland, Springer, 2021, p.10; Arkady Plotnitsky, Niels Bohr and Complementarity: an Introduction, New York, Springer, 2013, p.10.

2. Paula Gottlieb, "Aristotle on Non-Contradiction (Stanford Encyclopedia of Philosophy)," Stanford.edu, 2019; Laurence R. Horn, "Contradiction", (Stanford Encyclopedia of Philosophy), Edward N. Zalta (ed.), Winter 2018.

3. Mario Bunge, *Causality and Modern Science* (New Brunswick, N.J.: Transaction Publishers, 2009), p.33.

4. Philip Clayton and Paul Davies, *The Re-Emergence of Emergence : The Emergentist Hypothesis from Science to Religion* (Oxford: Oxford Univ. Press, 2008).

5. Aristotle, *Metaphysics*, trans. C.D.C. Reeve (Indianapolis, Cambridge Hackett Publishing Company, 2016), p.51.

The Application of Noncontradiction is No Longer an a Priori Certainty

1. Edwin Arthur Burtt, *The Metaphysical Foundations of Modern Physical Science* (New York: Harcourt, Brace & Company, Inc., 1925), p.72.

2. Edwin Arthur Burtt, *The Metaphysical Foundations of Modern Physical Science* (New York: Harcourt, Brace & Company, Inc., 1925), p.216.

3. Karin de Boer, Kant's Reform of Metaphysics: The Critique of Pure Reason Reconsidered, Cambridge, United Kingdom; New York, NY, Cambridge University Press, 2020, p.41.

4. Immanuel Kant, *Critique of Pure Reason* {A1781/B1787}, trans. Paul Guyer and Allen W. Wood (Cambridge: Cambridge Univ. Press, 1998), p.A133/B172/268.

Quantum Discontinuity is an Ontological Problem

1. Helge Kragh, *Niels Bohr and the Quantum Atom : The Bohr Model of Atomic Structure 1913-1925* (Oxford: Oxford University Press, 2012).

2. Paula Gottlieb, "Aristotle on Non-Contradiction (Stanford Encyclopedia of Philosophy)," Stanford.edu, 2019, https://plato.stanford.edu/entries/aristotle-noncontradiction/ .

3. Ferdinand de Saussure, *Course in General Linguistics* {1916}. The Philosophical Library, Inc., 1959.

4. Aristotle and Hugh Lawson-Tancred, *Metaphysics* (London: Penguin Books, 1998), p.80.85.

The Critical Ontology of Nicolai Hartmann

1. Herbert Spiegelberg, *The Phenomenological Movement : A Historical Introduction : Vol. 1-2.* (The Hague: Springer, 1960) p.358.

2. ibid., p.82

3. Maurice Merleau-Ponty, *Phenomenology of Perception* {1945}, trans. Donald A. Landes (New York. Routledge, 2012), p.lxxi.

4. Michael Landmann, "Nicolai Hartmann and Phenomenology," *Philosophy and Phenomenological Research* 3, no. 4 (June 1943): 393, https://doi.org/10.2307/2102844, p.395.

Kant's Unwitting Error

1. Immanuel Kant, *Prolegomena to Any Future Metaphysics That Will Be Able to Come Forward as Science* {1783}, ed. Karl Ameriks and Desmond M. Clarke, trans. Gary Hatfield (Cambridge Cambridge Univ. Press, 2004), p.28.

2. Arkady Plotnitsky, *The Principles of Quantum Theory, from Planck's Quanta to the Higgs Boson the Nature of Quantum Reality and the Spirit of Copenhagen* (Switzerland: Springer International Publishing, 2016), p.x.

3. Daniel N Robinson, *How Is Nature Possible? : Kant's Project in the First Critique* (London ; New York: Continuum, 2012).

4. Terry Pinkard, "Idealism," in *The Oxford Handbook of German Philosophy in the Nineteenth Century*, ed. Michael N. Foster and Kristin Gjesdal (Oxford University Press, 2015), p.3.

5. Keith R. Peterson, "An Introduction to Nicolai Hartmann's Critical Ontology," *Axiomathes* 22, no. 3 (February 29, 2012): 291–314, https://doi.org/10.1007/s10516-012-9184-1 , p.294.

6. Immanuel Kant, *Critique of Pure Reason* {A1781/B1787}, trans. Paul Guyer and Allen W. Wood (Cambridge: Cambridge Univ. Press, 1998), p.110 (Bxvii).

7. George Berkeley, *Principles of Human Knowledge and Three Dialogues* (Penguin UK, 1988).

8. Immanuel Kant, *Prolegomena to Any Future Metaphysics That Will Be Able to Come Forward as Science* {1783}, ed. Karl Ameriks and Desmond M. Clarke, trans. Gary Hatfield (Cambridge Cambridge Univ. Press, 2004).

9. Ibid., p.27.

10. Karin de Boer, *Kant's Reform of Metaphysics : The Critique of Pure Reason Reconsidered* (Cambridge, United Kingdom ; New York, Ny: Cambridge University Press, 2020), p.47.

Hegel Made the Same Unwitting Error

1. Will Durant, *The Story of Philosophy* (Garden City, N.Y. : Garden City Publishing Company, 1927), p.318.

2. Kierkegaard, Søren, and Lee M. Hollander. *Selections from the Writings of Kierkegaard.* (The University of Texas, Austin, 1923), p.19.

3. Hans-Georg Gadamer, *Heidegger's Ways* (Albany: State Univ. Of New York Press, 1994), p.3.

4. Willis F. Overton, "A New Paradigm for Developmental Science: Relationism and Relational-Developmental Systems," *Applied Developmental Science* 17, no. 2 (April 2013): 94–107; Carlo Rovelli, "Relational Quantum Mechanics," *International Journal of Theoretical Physics* 35, no. 8 (August 1996): 1637–78.

5. Nicolai Hartmann, "How Is Critical Ontology Possible? Toward the Foundation of the General Theory of the Categories, Part One {1923}," ed. Keith R. Peterson, *Axiomathes* 22, no. 3 (April 13, 2012): 315–54, https://doi.org/10.1007/s10516-012-9183-2 ., p.339.

6. Ibid., p.318.

Again, the Problem Starts in Western Metaphysics

1. Chuck Lorre and Stephen Molaro, "Young Sheldon S04E07," *Netflix*, 2020, S04,E07 6 minutes.

2. Thomas Reid, *An Inquiry into the Human Mind on the Principles of Common Sense* (Anderson and MacDowall, and James Robertson, Parliament Square, 1810), p.64.

BIBLIOGRAPHY

Aristotle. *Metaphysics*. Translated by C.D.C. Reeve. Indianapolis, Cambridge Hackett Publishing Company, 2016.

Aristotle. *Metaphysics*. Translated by Hugh Lawson-Tancred. London: Penguin Books, 1998.

Berkeley, George. *Principles of Human Knowledge and Three Dialogues*. Penguin UK, 1988.

Boer, Karin de. *Kant's Reform of Metaphysics : The Critique of Pure Reason Reconsidered*. Cambridge, United Kingdom ; New York, Ny: Cambridge University Press, 2020.

Bunge, Mario. *Causality and Modern Science*. New Brunswick, N.J.: Transaction Publishers, 2009.

Clayton, Philip, and Paul Davies. *The Re-Emergence of Emergence : The Emergentist Hypothesis from Science to Religion*. Oxford: Oxford Univ. Press, 2008.

Durant, Will. *The Story of Philosophy*. Garden City, N.Y. : Garden City Publishing Company, 1927.

Edwin Arthur Burtt. *The Metaphysical Foundations of Modern Physical Science*. New York: Harcourt, Brace & Company, Inc., 1925.

Gadamer, Hans-Georg. *Heidegger's Ways*. Albany: State Univ. Of New York Press, 1994.

Gottlieb, Paula. "Aristotle on Non-Contradiction (Stanford Encyclopedia of Philosophy)." Stanford.edu, 2019. .

Hartmann, Nicolai. "How Is Critical Ontology Possible? Toward the Foundation of the General Theory of the Categories, Part One {1923}." Edited by Keith R. Peterson. *Axiomathes* 22, no. 3 (April 13, 2012): 315–54. .

Heidegger, Martin. *Being and Time* {1927}. Translated by John Macquarrie and Edward Robinson. Blackwell, 2001.

Horn, Laurence R., 'Contradiction', The Stanford Encyclopedia of Philosophy, Edward N. Zalta (ed.), Winter 2018

Kant, Immanuel. *Critique of Pure Reason* {A1781/B1787}. Translated by Paul Guyer and Allen W. Wood. Cambridge: Cambridge Univ. Press, 1998.

Kant, Immanuel. *Prolegomena to Any Future Metaphysics That Will Be Able to Come Forward as Science* {1783}. Edited by Karl Ameriks and Desmond M. Clarke. Translated by Gary Hatfield. Cambridge Cambridge Univ. Press, 2004.

Kierkegaard, Søren, and Lee M. Hollander. *Selections from the Writings of Kierkegaard*. The University of Texas, Austin, 1923.

Lorre, Chuck, and Stephen Molaro. "Young Sheldon S04E07." Netflix, 2020.

Plotnitsky, Arkady. *Niels Bohr and Complementarity: an Introduction*. Springer, 2013.

Robinson, Daniel N. *How Is Nature Possible? : Kant's Project in the First Critique*. London ; New York: Continuum, 2012.

Saussure, Ferdinand de, et al. *Course in General Linguistics 1916*. The Philosophical Library, Inc., 1959.

Kragh, Helge. *Niels Bohr and the Quantum Atom : The Bohr Model of Atomic Structure 1913-1925*. Oxford: Oxford University Press, 2012.

Landmann, Michael. "Nicolai Hartmann and Phenomenology." *Philosophy and Phenomenological Research* 3, no. 4 (June 1943): 393. .

Maudin, Tim. "The Labyrinth of Quantum Logic." *History and Philosophy of Physics*, 2018.

Merleau-Ponty, Maurice. *Phenomenology of Perception* {1945}. Translated by Donald A. Landes. New York: Routledge, 2012.

Overton, Willis F. "A New Paradigm for Developmental Science: Relationism and Relational-Developmental Systems." *Applied Developmental Science* 17, no. 2 (April 2013): 94–107. .

Peterson, Keith R. "An Introduction to Nicolai Hartmann's Critical Ontology." *Axiomathes* 22, no. 3 (February 29, 2012): 291–314. .

Pinkard, Terry. "Idealism." In *The Oxford Handbook of German Philosophy in the Nineteenth Century*, edited by Michael N. Foster and Kristin Gjesdal. Oxford University Press, 2015.

Plotnitsky, Arkady. *The Principles of Quantum Theory, from Planck's Quanta to the Higgs Boson the Nature of Quantum Reality and the Spirit of Copenhagen*. Switzerland: Springer International Publishing, 2016.

Reid, Thomas. *An Inquiry into the Human Mind on the Principles of Common Sense*. Anderson and MacDowall, and Jame Robertson, Parliament Square, 1810.

Rovelli, Carlo. "Relational Quantum Mechanics." *International Journal of Theoretical Physics* 35, no. 8 (August 1996): 1637–78. .

Spiegelberg, Herbert. *The Phenomenological Movement : A Historical Introduction : Vol. 1-2*. The Hague: Springer, 1960.

www.ingramcontent.com/pod-product-compliance
Lightning Source LLC
Chambersburg PA
CBHW051451290426
44109CB00016B/1714